# Green Transportation and New Advances in Vehicle Routing Problems

Houda Derbel • Bassem Jarboui • Patrick Siarry
Editors

# Green Transportation and New Advances in Vehicle Routing Problems

 Springer

*Editors*
Houda Derbel
Department of Quantitative Methods
and CS
University of Carthage
Nabeul, Tunisia

Bassem Jarboui
Department of Business
Higher Colleges of Technology
Abu Dhabi, United Arab Emirates

Patrick Siarry
Laboratoire LiSSi (EA 3956)
Université Paris-Est Créteil
Créteil, France

ISBN 978-3-030-45314-5      ISBN 978-3-030-45312-1   (eBook)
https://doi.org/10.1007/978-3-030-45312-1

This Springer imprint is published by the registered company Springer Nature Switzerland AG.
The registered company address is: Gewerbestrasse 11, 6330 Cham, Switzerland

# Preface

The vehicle routing problem (VRP) is considered among the most challenging in combinatorial optimization. In recent years, the evolution of VRP toward achieving real-life applicable streams enables expedited VRP variants, models, solution techniques, and applications.

In supply chain management, in addition to minimizing transport costs, there is now critical attention paid to environmental aspects. This book provides new advances in vehicle routing problems with the means to develop green logistic systems. Although there is a variety of existing papers and studies on vehicle routing problems, the green remains a promising field that requires further study. To this end, we propose addressing both environmental aspects and vehicle routing problems (VRPs) together.

The interaction of classical variants of vehicle routing problems with green concerns can help organizations to overcome difficulties and to achieve sustainability goals while minimizing the total distance to serve customers by providing cost-effective solutions. For this purpose, different vehicle routing problems and their extensions taking into account operational constraints related to energy consumption are proposed. In response to these challenges, this book essentially addresses the trade-off between transportation and energy consumption. It presents new VRP models and solution approaches while considering green transportation.

The green vehicle routing problem (GVRP) is a variant of the standard vehicle routing problem designed to consider environmental externalities in addition to finding optimal routes. Interest in the green transportation area has considerably intensified in recent years. In this regard, this book "Green Transportation and New Advances in Vehicle Routing Problems" gives an overview of the green vehicle routing problem and focuses on the new trends in VRP motivated by achieving environmental objectives in transportation. It proposes new advances within vehicle routing models, new solution techniques for solving GVRP, and new applications toward green transportation.

The chapters in this book evolve new models and solution algorithms for solving different variants of vehicle routing problems that consider challenging environmental aspects. They present not only an up-to-date review of vehicle routing problems and practical perspectives but also solution approaches related to the green context to help organizations and companies achieve greener activities.

## Organization of the Book

This book is organized into 8 chapters.

The chapter "New Advances in Vehicle Routing Problems: A Literature Review to Explore the Future" by M. Asghari and S.M.J. Mirzapour Al-e-Hashem aims to provide a comprehensive and up-to-date review of the three major and applicable streams of vehicle routing problems (VRPs), namely the electric vehicle routing problem (EVRP), the green vehicle routing problem (GVRP), and the hybrid vehicle routing problem (HVRP), including formulation techniques, methods of solution, and areas of application. The main objective of this chapter was to organize the available literature so as to provide a reference point for future research on new aspects of routing problems.

The chapter "A Robust Optimization for a Home Healthcare Routing and Scheduling Problem Considering Greenhouse Gas Emissions and Stochastic Travel and Service Times" by A.M. Fathollahi-Fard, A. Ahmadi, and B. Karimi deals with home healthcare (HHC) services for elderly people in developed countries. A HHC system aims to consider a number of caregivers, starting with a pharmacy, serving a set of patients at their home with different home care services, including housekeeping, nursing, and physiotherapy. To meet the demands of patients in a timely fashion, the caregiver has to deal with some stochastic parameters to carry out a valid plan to visit the patients. The logistics activities of this system need efficient optimization models and algorithms to find a valid plan for caregivers addressing the HHC routing and scheduling problem (HHCRSP). A bi-objective robust optimization model is developed and solved using a set of efficient metaheuristic algorithms by considering the green dimensions of the logistic activities. The authors also apply the Keshtel algorithm (KA) to solve the problem and compare it with well-known algorithms based on different criteria.

The chapter "A Skewed General Variable Neighborhood Search Approach for Solving the Battery Swap Station Location-Routing Problem with Capacitated Electric Vehicles" by M. Affi, H. Derbel, B. Jarboui, and P. Siarry considers a new variant of location-routing problem for an electric vehicle with a single depot; an example of a location decision relates to the battery swap stations combined with considering the vehicle routing. The authors propose a solution method based on a skewed version of the variable neighborhood search (SVNS) metaheuristic in order to reduce the energy consumption and the pollution caused by transport operations. The results show that the SVNS is competitive with state-of-the-art methods.

The chapter "The Cumulative Capacitated Vehicle Routing Problem Including Priority Indexes" by K. Corona-Gutiérrez, M-L. Cruz, S. Nucamendi-Guillén, and E.O Benítez studies the cumulative capacitated vehicle routing problem, including priority indexes. This problem can be effectively implemented in commercial and public environments, where green concerns are incorporated. A mixed integer formulation is developed aiming at minimizing two objectives: the total latency and the total tardiness of the system and solved using the AUGMECON approach to obtain true efficient Pareto fronts. Moreover, two metaheuristics were developed to solve the problem, one based on the non-dominated sorting genetic algorithm (NSGA) and the other based on particle swarm optimization (PSO). The authors show that the proposed model and algorithms can be useful to solve a wide variety of situations, where economic, environmental, and social concerns are involved.

The chapter "Solution of a Real-Life Vehicle Routing Problem with Meal Breaks and Shifts" by Ç. Karademir, Ü. Bilge, N. Aras, G.B. Akkuş, G. Öznergiz, and O. Doğan addresses an extension of the classical vehicle routing problem with time windows (VRPTW). This variant incorporates shifts as well as meal breaks of the delivery personnel and requires the waiting time of a vehicle occurring between the customer arrival time and the service start time to be less than a certain threshold value, which gives rise to the VRPTW including breaks and shifts (VRPTW-BS). Three methods are developed and applied on instances of different sizes to solve the problem.

The chapter "A Decomposition-Based Heuristic for a Waste Cooking Oil Collection Problem" by C. Gultekin, O.B. Olmez, B. Balcik, A. Ekici, and O.O. Ozener focuses on designing a waste cooking oil (WCO) collection network motivated by the oil collection operations of a biodiesel company, which uses the collected WCO as a raw material in biodiesel production. The network involves a biodiesel facility, a set of collection centers (CCs), and source points (SPs) each of which represents a group of households. WCO is produced by households and commercial organizations, which poses a serious threat to the environment if disposed improperly. The problem addressed in this chapter is a new variant of the secondary facility location-routing problem, where the bin allocation decision is also under consideration. The results of the mixed-integer programming (MIP) model and the decomposition-based heuristic (DA) with respect to both solution quality and computational time are given and compared.

The chapter "Time-Dependent Green Vehicle Routing Problem" by G. Masghati-Amoli and A. Haghani describes a model for a time-dependent green vehicle routing problem (TDGVRP) with a mixed fleet of electric (ECV) and internal combustion engine commercial vehicles (ICCV) that finds the optimal fleet design and routes for last-mile delivery operations. The GVRP model proposed in this chapter takes into account the limitations in the adoption of both ECVs and ICCVs. A mathematical formulation and a heuristic algorithm are proposed for the defined TDGVRP, and numerical experiments are designed and tested to demonstrate their capabilities. The results show that the proposed heuristic is efficient in finding sound solutions and can be used to identify the changes in fleet design and routing of last-mile delivery operations as a result of green logistics policies.

The chapter "Recent Developments in Real-Life Vehicle Routing Problem Applications" by L. Simeonova, N. Wassan, N. Wassan, and S. Salhi discusses some of the main vehicle routing problem categories motivated by real operations, namely integrated problems, problems with alternative objective functions, and those problems that are highly problem specific and have unique characteristics. It provides a discussion on the evolution of real-life routing problems and evaluates some important methodological considerations and aspects, which are typically overlooked in the literature, such as the role of data, methodological comparability, implementation, and benefits realization. Finally, this chapter offers the guidance for best practice and future research in the area of real-life routing problems.

## *Audience*

This book is dedicated to master's and Ph.D. students, researchers in operation research, especially in the green vehicle routing field, and professionals in industry. They will find meaningful models and solution approaches that help them develop sustainable operations.

Nabeul, Tunisia                                                    Houda Derbel
Abu Dhabi, United Arab Emirates                      Bassem Jarboui
Créteil, France                                                    Patrick Siarry

# Contents

1 New Advances in Vehicle Routing Problems: A Literature
  Review to Explore the Future................................................ 1
  Mohammad Asghari and S. Mohammad J. Mirzapour Al-e-hashem

2 A Robust Optimization for a Home Healthcare Routing and
  Scheduling Problem Considering Greenhouse Gas Emissions
  and Stochastic Travel and Service Times................................. 43
  Amir Mohammad Fathollahi-Fard, Abbas Ahmadi,
  and Behrooz Karimi

3 A Skewed General Variable Neighborhood Search Approach for
  Solving the Battery Swap Station Location-Routing Problem
  with Capacitated Electric Vehicles....................................... 75
  Mannoubia Affi, Houda Derbel, Bassem Jarboui, and Patrick Siarry

4 The Cumulative Capacitated Vehicle Routing Problem
  Including Priority Indexes................................................ 91
  Karina Corona-Gutiérrez, Maria-Luisa Cruz,
  Samuel Nucamendi-Guillén, and Elias Olivares-Benitez

5 Solution of a Real-Life Vehicle Routing Problem with Meal
  Breaks and Shifts ........................................................ 131
  Çiğdem Karademir, Ümit Bilge, Necati Aras, Gökay Burak Akkuş,
  Göksu Öznergiz, and Onur Doğan

6 A Decomposition-Based Heuristic for a Waste Cooking Oil
  Collection Problem....................................................... 159
  Ceren Gultekin, Omer Berk Olmez, Burcu Balcik, Ali Ekici,
  and Okan Orsan Ozener

7   **Time-Dependent Green Vehicle Routing Problem** ....................... 177
    Golnush Masghati-Amoli and Ali Haghani

8   **Recent Developments in Real Life Vehicle Routing Problem
    Applications** ............................................................... 213
    Lina Simeonova, Niaz Wassan, Naveed Wassan, and Said Salhi

# List of Contributors

**Mannoubia Affi** Faculty of Economics and Management Sciences of Sfax, Sfax, Tunisia

**Abbas Ahmadi** Department of Industrial Engineering and Management Systems, Amirkabir University of Technology (Tehran Polytechnic), Tehran, Iran

**Gökay Burak Akkuş** Ekol Logistics, İstanbul, Turkey

**Necati Aras** Dept. of Industrial Engineering, Boğaziçi University, İstanbul, Turkey

**Mohammad Asghari** Department of Industrial Engineering and Management Systems, Amirkabir University of Technology (Tehran Polytechnic), Tehran, Iran

**Elías Olivares Benítez** Universidad Panamericana, Facultad de Ingeniería, Mexico, Mexico

**Ümit Bilge** Dept. of Industrial Engineering, Boğaziçi University, İstanbul, Turkey

**Karina Corona-Gutiérrez** Universidad Panamericana, Facultad de Ingeniería, Mexico, Mexico

**Maria-Luisa Cruz** Universidad Panamericana, Facultad de Ingeniería, Mexico, Mexico

**Carlos Delgado** Universidade Nova de Lisboa IMS, Lisbon, Portugal

**Houda Derbel** Faculty of Economics and Management Sciences of Sfax, Sfax, Tunisia

**Onur Doğan** Ekol Logistics, İstanbul, Turkey

**Amir Mohammad Fathollahi-Fard** Department of Industrial Engineering and Management Systems, Amirkabir University of Technology (Tehran Polytechnic), Tehran, Iran

**Ali Haghani** University of Maryland, College Park, MD, USA

**Moritz Hildemann** Universidade Nova de Lisboa IMS, Lisbon, Portugal

**Bassem Jarboui** Emirates College of Technology, Abu Dhabi, United Arab Emirates

**Çiğdem Karademir** Dept. of Industrial Engineering, Boğaziçi University, İstanbul, Turkey

**Behrooz Karimi** Department of Industrial Engineering and Management Systems, Amirkabir University of Technology (Tehran Polytechnic), Tehran, Iran

**Golnush Masghati-Amoli** University of Maryland, College Park, MD, USA

**S. Mohammad J. Mirzapour Al-e-hashem** Department of Industrial Engineering and Management Systems, Amirkabir University of Technology (Tehran Polytechnic), Tehran, Iran

**Samuel Nucamendi-Guillén** Universidad Panamericana, Facultad de Ingeniería, Mexico, Mexico

**Göksu Öznergiz** Ekol Logistics, İstanbul, Turkey

**Said Salhi** Centre for Logistics and Heuristic Optimisation (CLHO), Kent Business School, University of Kent, Canterbury, UK

**Patrick Siarry** Université Paris-Est Créteil Val-de-Marne, Laboratoire LiSSi, Créteil, France

**Lina Simeonova** Centre for Logistics and Heuristic Optimisation (CLHO), Kent Business School, University of Kent, Canterbury, UK

**Niaz Wassan** Centre for Logistics and Heuristic Optimisation (CLHO), Kent Business School, University of Kent, Canterbury, UK

**Naveed Wassan** Sukkur IBA University, Sukkur, Pakistan

# Acronyms

| | |
|---|---|
| AFV | Alternative fuel vehicle |
| AFVRP | Alternative fuel vehicle routing problem |
| ACO | Ant colony optimization |
| AFS | Alternative fuel station |
| ANFIS | Adaptive neuro-fuzzy inference system |
| BA | Bat algorithm |
| BSS-EV-LRP | Battery swap station location-routing problem with capacitated electric vehicles |
| BSS | Battery swap station |
| CG | Column generation |
| CS | Cuckoo search |
| CVRP | Capacitated vehicle routing problem |
| CMEM | Comprehensive modal emission model |
| CRBI | Cost ratio-based insertion |
| CAA | Civil Aviation Authority |
| DP | Dynamic programming |
| DEA | Differential evolution algorithm |
| EVRP | Electric vehicle routing problem |
| EV | Electric vehicle |
| EC | Function of electricity consumption |
| E-VNS | Evolutionary variable neighborhood search |
| EVRPTW | Electric vehicle routing problem with time window |
| ECR | Energy/electricity consumption rate |
| ETSPTW | Electric traveling salesman problem with time window |
| EMVRP | Energy minimizing vehicle routing problem |
| EVTOL | Electrical vertical take-off and landing technology |
| FAA | Federal Aviation Administration |
| FA | Firefly algorithm |
| GA | Genetic algorithm |
| GVNS | General variable neighborhood search |
| GHG | Greenhouse gas |

| | |
|---|---|
| GVRP | Green vehicle routing problem |
| GNA | Green navigation algorithm |
| HHCRSP | Home healthcare routing and scheduling problem |
| HSA | Harmony search algorithm |
| HVRP | Hybrid vehicle routing problem |
| ICEV | Internal combustion engine vehicle |
| ILS | Iterated local search |
| IDW | Inversed distance weight |
| ICCV | Internal combustion engine commercial vehicle |
| KA | Keshtel algorithm |
| LNS | Large neighborhood search |
| LS | Local search |
| LiDAR | Light detection and ranging |
| LRP | Location routing problem |
| LEZ | Low emission zone |
| MA | Memetic algorithm |
| MOPSO | Multi-objective particle swarm optimization |
| MOKA | Multi-objective Keshtel algorithm |
| MILP | Mixed-integer linear programming |
| NSGA-II | Non-dominated sorting genetic algorithm |
| NP-hard | Non-polynomial hard |
| NPS | Number of Pareto solutions |
| N | Nonlinear recharging |
| NS | Not specified properties |
| NRGA | Non-dominated ranking genetic algorithm |
| NSGA-II | Non-dominated sorting genetic algorithm II |
| PBA | Path-based approach |
| PSO | Particle swarm optimization |
| PRP | Pollution routing problem |
| PHEV | Plug-in hybrid electric vehicle |
| RSM | Response surface method |
| SDSS | Spatial decision support system |
| SA | Simulated annealing |
| SSA | Salp swarm algorithm |
| SNS | Spread of non-dominance solutions |
| SOC | State of charge |
| TDGVRP | Time-dependent green vehicle routing problem |
| TS | Tabu search |
| TPH | Two-phase heuristic |
| UAS | Unmanned aerial systems |
| VNS | Variable neighborhood search |
| VRP | Vehicle routing problem |
| VRPTW-BS | Vehicle routing problem time windows including meal breaks and shifts |
| VRPTWLB | Vehicle routing problem with time windows and lunch breaks |

# Chapter 1
# New Advances in Vehicle Routing Problems: A Literature Review to Explore the Future

**Mohammad Asghari and S. Mohammad J. Mirzapour Al-e-hashem**

**Abstract** The vehicle routing problem (VRP) is a combinatorial optimization problem that involves finding the optimal design of routes for a fleet of vehicles to serve the demands of a set of customers. This chapter investigates the evolution of three major and applicable streams of the VRPs, namely, electric vehicle routing problem (EVRP), green vehicle routing problem (GVRP), and hybrid vehicle routing problem (HVRP). This survey chapter aims to provide a comprehensive and structured overview of the state of knowledge and discuss important characteristics of the problems, including formulation techniques, methods of solution, and areas of application. We then show how they are different from the traditional vehicle routing problem. Finally, we present a summary table for each variant to emphasize some key features that represent the developments direction of researches.

## 1.1 Introduction

According to Nakata [144], the future estimation of energy systems indicates that a serious loss will happen by 2040. This future decline in energy availability and the pollution generated by unnatural resources lead to many efforts for preventing the damages by overcoming the pollution resulting from fossil fuel consumption. With technological developments, scientists concentrate on reducing carbon emissions by considering electric or hybrid vehicles as the base of their researches.

Using batteries is the main feature of an Electric Vehicle (EV) which directly influences the performance and cost of the driving [171]. Since the hybrid and electric vehicles are important for reducing environmental damage, researchers have been examining them even though they are still in limited use [82]. As we know, the research efforts before 2010 mainly focused on the traditional VRP. To study more details on the variants of traditional Vehicle Routing Problem (VRP) and

M. Asghari (✉) · S. M. J. Mirzapour Al-e-hashem
Department of Industrial Engineering and Management Systems, Amirkabir University
of Technology (Tehran Polytechnic), Tehran, Iran
e-mail: mohammad.asghari@aut.ac.ir; mirzapour@aut.ac.ir

© Springer Nature Switzerland AG 2020
H. Derbel et al. (eds.), *Green Transportation and New Advances in Vehicle Routing Problems*, https://doi.org/10.1007/978-3-030-45312-1_1

their features, refer to [21, 142]. A few studies on green routing problem especially EV had been conducted during this period. After 2010, routing problems covering energy consumption and pollution emissions started to draw close attention for researchers and became a hot topic in recent years.

We have examined and analyzed the articles in three major and applicable streams, namely, electric vehicle routing problem (EVRP), green vehicle routing problem (GVRP), and hybrid vehicle routing problem (HVRP). As shown in Table 1.1, about 236 papers are reviewed in this chapter that reflects the environmental sensitivity consideration in routing problems and one can observe that this concern is increasing day by day.

The chapter is organized as follows. Section 1.2 is devoted to a summary of recent publications in EVRP, while Sect. 1.3 contains the review on GVRP. This is

**Table 1.1** The scientific contribution reviewed in this chapter

| Year | Number of papers | The list of the papers |
|------|------------------|------------------------|
| 1976 | 1 | [35] |
| 1998 | 1 | [12] |
| 1999 | 1 | [50] |
| 2000 | 2 | [144, 165] |
| 2002 | 1 | [161] |
| 2004 | 1 | [176] |
| 2005 | 2 | [13, 115] |
| 2006 | 1 | [67] |
| 2007 | 6 | [33, 117, 137, 149, 179, 186] |
| 2008 | 6 | [8, 72, 74, 135, 145, 198] |
| 2009 | 3 | [20, 190, 203] |
| 2010 | 8 | [6, 15, 51, 58, 88, 126, 131, 187] |
| 2011 | 8 | [17, 39, 52, 73, 160, 194, 202, 217] |
| 2012 | 8 | [41, 48, 83, 85, 102, 175, 213, 215] |
| 2013 | 9 | [10, 54, 60, 61, 127, 143, 150, 205, 208] |
| 2014 | 20 | [2, 3, 38, 42, 43, 53, 63, 77, 87, 89, 95, 98, 112, 148, 151, 153, 164, 177, 184, 227] |
| 2015 | 21 | [5, 9, 24, 25, 36, 45, 68, 70, 71, 100, 109, 111, 122, 123, 142, 185, 192, 200, 214, 218] |
| 2016 | 34 | [4, 16, 18, 21, 26, 31, 34, 40, 44, 59, 69, 76, 78, 79, 82, 90, 94, 103, 108, 113, 124, 140, 146, 154, 158, 166, 167, 171, 174, 193, 195, 199, 206, 212, 221, 222] |
| 2017 | 39 | [1, 7, 11, 19, 27, 28, 32, 37, 49, 57, 62, 64, 65, 81, 84, 101, 110, 120, 132, 134, 139, 141, 155, 156, 159, 168, 173, 178, 180, 181, 188, 191, 197, 201, 219, 220, 226, 228, 235] |
| 2018 | 34 | [22, 29, 46, 55, 91, 91, 96, 97, 99, 104, 105, 114, 116, 119, 130, 133, 136, 138, 147, 152, 169, 182, 183, 189, 196, 207, 209, 229–231, 236] |
| 2019 | 29 | [14, 23, 30, 47, 66, 75, 80, 86, 93, 106, 107, 118, 121, 125, 128, 129, 157, 163, 170, 172, 204, 210, 211, 216, 224, 225, 232–234] |

followed by the surveys on HVRP in Sect. 1.4. To make this easier for interested readers, a table of summary on mode of formulation, methods of solution, and other essential features is given in each review section. Eventually, Sect. 1.5 provides the main findings of the literature review and directions for further research.


## 1.2   Electric Vehicle Routing Problem

In the last decade, EVs have been proposed as alternative-energy transportation for personal and commercial uses, and their market share has grown rapidly [54]. The environmentally friendly aspects of EVs, such as low noise pollution, no greenhouse gas emissions, and high-energy efficiency is the main reason of increase in their popularity that benefits logistics firms in achieving a green image from a growing number of environmentally and socially knowledgeable clients [44]. Moreover, state regulatory support is an influential factor that encouraged using EVs in both individual travel and public transportations [103]. Figure 1.1 displays the obvious increase experienced during the last years in the number of EV-related articles published in Scopus-indexed journals, which proves the growing interest that the use of EVs in routing problems is arising among researchers and practitioners. Readers can find recent sustainability investigation and realistic evaluation techniques of life cycle carbon emission generated by electric vehicles in [31, 55]. Along with the environmental advantages, these kinds of vehicles are considered attractive than conventional fossil fuel-powered vehicles in terms of a cost perspective. In

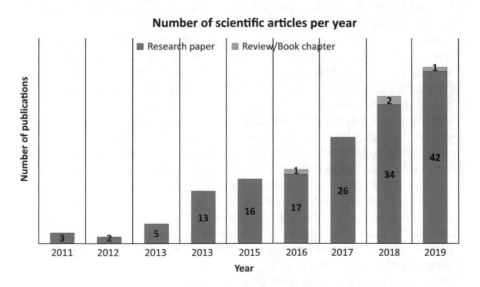

**Fig. 1.1** Evolution of EVRP related publications in Scopus-indexed journals

comparison to the cost of petrol/gasoline consumed by a conventional vehicle, when traversing the same distance, an EV costs 85–90% less.

Despite comprehensive researches on the characteristics and attributes of EVs and the design of charging infrastructure, the network modeling of EVs is still developing and limited. Because of the assuring opportunity of the EVs for reducing pollution impacts and transportation costs compared with fossil fuel based engines, the EVRP, which deals with planning routes for commercial EVs, is one of the logistics distribution/service field investigated in the recent years by researchers [29, 80, 113, 153].

Table 1.2 classifies the main features of the studies done on EVRP as an extension to the review provided by Keskin et al. [93]. The headings of the column represent recharging function (Rch), the function of electricity consumption (EC), fleet type, charger composition in the stations, charge amount, considering time windows (TW), determining the location of station (Lc), objective function (Obj.), and solution approach; linear and nonlinear recharging time functions are indicated by L and N; homogeneous and heterogeneous fleets are indicated by Ho and Ht; NS, *, **, ***, **** stand for not specified properties, consumption depends on speed, mass and gradient of the terrain, linear charging with different cost components for different charge levels, linear charging with different cost and rates for different chargers, linear and constant charging for battery swapping and parallel with considering service time, respectively; A-, G-, AC, CG, DP, E-VNS, HC, ILS, PBA, PSO, TS, and WSM correspond to adaptive-, general-, ant colony, column generation, dynamic programming, evolutionary variable neighborhood search, heuristic concentration, iterated local search, path-based approach, particle swarm optimization, tabu search, and weighted sum method, respectively. CS is also used for the articles that just apply a commercial solver for reporting results. Table 1.3 defines the encoding used for each objective function.

The main difficulty associated with planning the EVs is how to include the visiting stations into the optimization, because the EVs have short travel range and their recharging can be time-consuming and uncomfortable during the travel. Some of the studies incorporate recharging at the stations during the travel, such as in the green routing problems [7, 48, 53], partial charging EVRP [28, 90, 182], battery swapping EVRP [81, 86, 218], EVRP with a mix of partial charging and battery swapping [86, 152, 181], and the EVRP considering a nonlinear recharging time function [64–66, 141]. Macrina et al. [128] present a table providing most of the recent mathematical models that use linear distance-dependent energy consumption.

Chargers in the stations may be equipped with different options including power voltage and maximum current, which have an effect on the duration of charging. Some models take stations with different chargers into account [53, 91, 111, 177]. Additionally, some of researches considered the heterogeneous fleet composed of various EVs with different features [44] whereas other researches cover different internal combustion engine vehicles (ICEVs) as well [71, 80, 97, 177]. Recently several studies have incorporated the location of the stations in the routing optimization [111, 152, 181, 218].

**Table 1.2** EVRP literature overview [93]

| Papers | Rch | | EC | | Fleet type | | Charger composition | | Charging | | TW | Lc | Obj. | Solution method |
|---|---|---|---|---|---|---|---|---|---|---|---|---|---|---|
| | L | N | L | N | Ho | Ht | Single | Multi. | Full | Partial | | | | |
| [39] | | | ✓ | | EVs | | ✓ | | ✓ | ✓ | | | 1, 2, 8, 15 | Iterative construction, improvement heuristics |
| [48] | | | ✓ | | AFVs | | ✓ | | ✓ | | | | 2 | Several heuristics |
| [53] | ✓ | | ✓ | | EVs | | | ✓ | | ✓ | ✓ | | 2, 15 | Several heuristics and SA |
| [177] | ✓ | | ✓ | | | EVs, ICEVs | | ✓ | | ✓ | ✓ | | 1, 2, 15 | Several heuristics |
| [184] | ✓ | | ✓ | | EVs | | ✓ | | ✓ | | ✓ | | 1, 2 | Hybrid VNS and TS |
| [9] | ✓ | | ✓ | | PHEVs | | ✓ | | | ✓ | | | 6, 9, 10 | DP |
| [25] | ✓ | | ✓ | | EVs | | ✓ | | | ✓ | ✓ | | 1, 2, 4, 15 | VNS branching |
| [45] | ✓ | | ✓ | | EVs | | ✓ | | | ✓ | ✓ | | 2 | Hybrid VNS and TS |
| [71] | *** | | * | | | EVs, ICEVs | ✓ | | | ✓ | ✓ | | 2, 8, 9, 10, 15 | ALNS |
| [111] | *** | | ✓ | | EVs | | | ✓ | ✓ | ✓ | ✓ | ✓ | 5, 8, 15 | Hybrid AVNS and TS |
| [218] | | | ✓ | | EVs | | BSS | | | | | ✓ | 2, 5 | Several heuristics |
| [44] | ✓ | | ✓ | | EVs | | ✓ | | ✓ | ✓ | ✓ | | 2 | Branch-price-and-cut |
| [76] | ✓ | | ✓ | | EVs | | ✓ | | | ✓ | ✓ | | 1, 2, 15 | WSM |
| [79] | ✓ | | ✓ | | | EVs | ✓ | | ✓ | ✓ | ✓ | | 1, 2 | Branch-price, and ALNS |
| [90] | ✓ | | ✓ | | EVs | | ✓ | | | ✓ | | | 1, 2 | ALNS |
| [94] | | | ✓ | | AFVs | | ✓ | | ✓ | | | | 2 | Branch-and-cut |
| [108] | | | NS | | EVs | | BSS | | | | | | 8 | DP |
| [113] | ✓ | | ✓ | | EVs | | ✓ | | ✓ | | | | 1, 2, 15 | CS |
| [140] | | | ✓ | | AFVs | | ✓ | | ✓ | | | | 2 | MSH |
| [174] | ✓ | | ✓ | | EVs | | ✓ | | ✓ | ✓ | ✓ | | 2 | GVNS, DP |
| [212] | ✓ | | ✓ | | EVs | | ✓ | | | ✓ | ✓ | | 1, 2 | ALNS |
| [7] | | | ✓ | | AFVs | | ✓ | | ✓ | | | | 2 | An exact algorithm |
| [27] | ✓ | | ✓ | | EVs | | ✓ | | | ✓ | ✓ | | 1, 8 | VNSB based matheuristic |
| [64] | | ✓ | ✓ | | EVs | | ✓ | | | ✓ | | | 8 | Two-stage matheuristic |
| [81] | | | ✓ | | EVs | | BSS | | | | | ✓ | 2, 5 | AVNS |

(continued)

**Table 1.1** (continued)

| Papers | Rch L | Rch N | EC L | EC N | Fleet type Ho | Fleet type Ht | Charger composition Single | Charger composition Multi. | Charging Full | Charging Partial | TW | Lc | Obj. | Solution method |
|---|---|---|---|---|---|---|---|---|---|---|---|---|---|---|
| [101] | ✓ | | ✓ | | AFVs | | ✓ | | | ✓ | | | 2 | CS |
| [141] | | ✓ | ✓ | | EVs | | | ✓ | | ✓ | | | 8 | Hybrid ILS and HC |
| [181] | ✓ | | ✓ | | EVs | | ✓ | | ✓ | ✓ | ✓ | ✓ | 1, 2, 5 | CS |
| [197] | ** | | ✓ | | EVs | | | ✓ | | ✓ | | | 2, 4, 15 | Several exact and heuristic methods |
| [22] | ✓ | | ✓ | | EVs | | ✓ | | ✓ | | | | 1, 2 | LNS, Exact algorithm |
| [29] | | | ✓ | | AFVs | | ✓ | | ✓ | | | | 2 | An exact method |
| [91] | ✓ | | ✓ | | EVs | | | ✓ | | ✓ | ✓ | | 1, 15 | ALNS based matheuristic |
| [96] | | ✓ | ✓ | | EVs | | | ✓ | | ✓ | | ✓ | 5, 8 | ALNS based matheuristic |
| [97] | ✓ | | | ✓ | | EVs, ICEVs | ✓ | | | ✓ | ✓ | | 11 | CS |
| [99] | ✓ | ✓ | ✓ | | EVs | | ✓ | | | ✓ | | | 7 | Several exact and heuristic methods |
| [104] | ✓ | | ✓ | | EVs | | Single chargers & WCS | | | ✓ | | | 8, 15 | CS |
| [130] | | | ✓ | | AFVs | | ✓ | | ✓ | | ✓ | | 2 | CS |
| [136] | **** | | * | | | EVs | BSS | | | ✓ | ✓ | | 2 | E-VNS |
| [152] | **** | | ✓ | | EVs | | BSS | | ✓ | | ✓ | ✓ | 1, 2, 5 | CS |
| [182] | ✓ | | ✓ | | EVs | | ✓ | | | ✓ | ✓ | ✓ | 1, 5, 8 | ALNS, DP |
| [207] | | ✓ | ✓ | | EVs, ICEVs | | | ✓ | ✓ | | ✓ | | 1, 2, 6, 15 | GRASP based matheuristic |
| [209] | | | ✓ | | EVs | | BSS | | ✓ | | ✓ | | 12 | Heuristic procedures |
| [230] | ✓ | | ✓ | | EVs | | ✓ | | | ✓ | ✓ | | 1, 2, 4, 8, 15 | TS |
| [14] | ✓ | | * | | EVs | | ✓ | | ✓ | | ✓ | | 3 | Bellman–Ford algorithm |
| [30] | | | ✓ | | AFVs | | ✓ | | ✓ | | | | 2 | PBA |
| [66] | | ✓ | ✓ | | EVs | | ✓ | | | ✓ | | | 8 | CS |
| [80] | ✓ | | ✓ | | | EVs, ICEVs, PHEVs | ✓ | | | ✓ | ✓ | | 1, 9, 15 | GA, LNS with IP solver |
| [86] | | | ✓ | | | EVs | BSS | | | ✓ | | | 2, 13, 14 | Hybrid CG and ALNS |
| [128, 129] | ✓ | | ✓ | | EVs, ICEVs | | ✓ | | | ✓ | ✓ | | 2, 15 | ILS |
| [157] | | | ✓ | | EVs | | ✓ | | | | | | 16 | Two-phase based LNS |

**Table 1.3** Explanation of the objective functions used in Tables 1.2 and 1.5

| | |
|---|---|
| 1: Vehicle cost/wage | 9: Battery cost |
| 2: Total distance/travel cost | 10: Charging cost |
| 3: Fuel consumption/emission | 11: Profit/revenue |
| 4: Waiting cost | 12: Operational costs |
| 5: Station installation cost | 13: Battery swapping cost |
| 6: Stopping cost at a station | 14: Unit time penalty for violated time windows |
| 7: Total makespan (duration) | 15: Recharging cost |
| 8: Fuel cost | 16: Fixed and maintenance cost |

In the majority of papers, recharging is done at special charging stations, although several studies investigate replacing discharged battery with a full charged one which is called battery-exchange stations [86, 108, 136, 152, 209], and recharging the EV by an inductive charging system located along the roads which is known as wireless charging systems [104]. Recently, some papers have studied the context of the two-echelon EVRP [22, 86], reverse logistics [231], technician routing [207], and pickup and delivery problem [76, 130]. A comprehensive review of the goods distribution with EVs can be found in [154, 155].

Conrad and Figliozzi [39] presented the first research on the EVRP where EVs can be recharged at some client places to continue their trip. The model had a fixed charging time by allowing fully charge or maximize 80% of the energy capacity at just client places. Since then, several extensions have been proposed. In the EVRP, aside from battery capacity, other factors such as considering time windows for customers and load capacities for vehicles are repeatedly taken into account. Conrad and Figliozzi suggested a bi-objective mathematical formulation that considers EVs. The objectives aim to minimize the total traveling distance and the number of routes. To solve the proposed problem, the authors offered an iterative route construction and improvement algorithm. Worley et al. [213] investigated a mixed-integer mathematical formulation for the EVRP with charging station siting constraints to concurrently minimize the total cost of traveling, recharging, and station location. Wang and Cheu [208] studied the recharging operations of a fleet of electric taxis. Then, to minimize the total distance traveled and maximum route time, they proposed a TS algorithm.

Currently, a few related papers in the context of EVs have addressed the location-routing problem in an uncertain environment. For example, a robust optimization structure was applied by Schiffer and Walther [182]. Their study analyzes uncertain parameters consisting of demand, customer locations, and time windows in a simultaneous problem of locating charging stations and routing EVs. In this context, Lu et al. [114] suggested a multi-objective mathematical model that considers load dispatch of a microgrid accessing stochastically to a large-sized of uncoordinated charging of EVs. The objective functions aim to minimize the operating cost, pollutant treatment cost, and load variance. The authors proposed an enhanced PSO algorithm to solve the problem effectively. Under three different scheduling scenarios, the dispatch results showed shifting charging loads from high-priced periods to low-priced periods in the presence of the coordinated charging mode of EVs.

Regarding the uncertain service time spent at the stations, Sweda et al. [197] investigated the stations under the probability unavailability. When an EV reaches an occupied station, it may need to wait for a random time to get service. A relevant problem was solved by Froger et al. [64] who determined an inadequate number of chargers for each station and a vehicle has to wait for recharging while other EVs in the fleet are using the chargers. Although in the proposed problem, the routing and charging features affect the use of the chargers, the queue lengths do not depend on these decisions. The waiting times at the alternative-fuel stations (AFSs) were recently studied by Bruglieri et al. [29]. The authors planned the routing of the alternative-fuel powered vehicles (AFVs) so that the consumed times for refueling of the AFVs do not overlap in the fuel stations. Finally, an exact approach is used to minimize the total distance traveled. Kullman et al. [99] considered a public–private recharging policy for the EVRP in which a charger in the depot belongs to the company, while the vehicles can also be recharged at public stations. If the EVs recharge at public stations, they may queue using an $M/M/\psi_c/\infty$ queuing system, otherwise use the depot charger as soon as they arrive. The results illustrated a substantial decrease in the cost of public–private strategies compared to private-only strategies.

Considering the literature, many different strategies exist for solving the EVRP and its variations. In a strong sense, the EVRP can equally be recognized as NP-hard problem because the EVRP is a generalization of the VRP, which is known as an NP-hard [44, 230]. Hence, metaheuristic optimization algorithms, which are earlier introduced for global optimization problems, are applied in order to solve most of the studies in this field. Table 1.4 reviews the metaheuristic algorithms-based solution approaches employed in the EVRP context. To improve algorithm efficiency, the integrated approaches, such as column generation, metaheuristic, dynamic programming, etc., have been applied as a subroutine in the algorithms. It is worth noting that most of the papers considered large neighborhood search and population-based algorithms like genetic algorithm, although a permutation order-based coding system can allow rendering a solution without doing any transformation.

Schneider et al. [184] done the first attempts for considering the EVRP with time windows, referred to as EVRPTW for simplicity, in which time consumed for recharging depends on the amount of energy left upon arrival at a recharging vertex. Recharging depends on the battery state of charge (SOC). It continues until the battery capacity is fully charged, which its duration is assumed to be commensurate with the amount of the energy transferred. The authors used a hierarchical objective, which minimizes the number of vehicles first and the total traveled distance second. To solve the proposed EVRPTW, a hybrid metaheuristic VNS/TS algorithm is developed and its performance is examined by several benchmark data sets related to VRPs. Goeke and Schneider [71] extended the same problem which incorporates a mixed fleet of EVs and ICEVs. The objective function aims to minimize the

**Table 1.4** Metaheuristic optimization algorithms employed in the EVRP

| Study | ACO | CS | DEA | GA | ILS | A-/LNS | MA | SA | A-/TS | A-/E-/VNS | Combined with |
|---|---|---|---|---|---|---|---|---|---|---|---|
| [208] | | | | | | | | | ✓ | | |
| [53] | | | | | | | | ✓ | | | |
| [148] | | | | ✓ | | | | | | | TS |
| [164] | | | | | | | | ✓ | | | |
| [184] | | | | | | | | | ✓ | | TS |
| [5] | | | | ✓ | | | | | | | TS |
| [24, 25] | | | | | | | | | ✓ | | Matheuristic |
| [45] | | | | | | | | | ✓ | | TS |
| [71] | | | | | ✓ | | | | | | |
| [111] | | | | | | | | | ✓ | | TS |
| [218] | | | | ✓ | | | | | | | Expert knowledge |
| [78] | | | | | | ✓ | | | | | Dynamic programming |
| [79] | | | | | | ✓ | | | | | Dynamic programming |
| [90] | | | | | | ✓ | | | | | |
| [158] | | | | ✓ | | | | | | | Set partitioning |
| [174] | | | | | | | | | | ✓ | Dynamic programming |
| [212] | | | | | ✓ | | | | | | |
| [1] | | | | ✓ | | | | | | | Markov decision process |
| [11] | | ✓ | | | | | | | | | |
| [28] | | | | | | | | | | ✓ | Matheuristic |
| [81] | | | | | | | | | ✓ | | |
| [141] | | | | ✓ | | | | | | | Heuristic concentration |
| [173] | | | | | | | ✓ | | | | GA |
| [188] | | | | ✓ | | | | | | | Dynamic Dijkstra |
| [235] | | | | ✓ | | | | | | | |
| [22] | | | | | ✓ | | | | | | |
| [116] | | | | | ✓ | | | | | | |
| [91] | | | | | ✓ | | | | | | |
| [92] | | | | | ✓ | | | | | | Matheuristic |
| [96] | | | | | ✓ | | | | | | Matheuristic |
| [136] | | | | | | | | | | ✓ | |
| [183] | | | | | | ✓ | | | | | Dynamic programming |

(continued)

**Table 1.4** (continued)

| Study | ACO | CS | DEA | GA | ILS | A-/LNS | MA | SA | A-/TS | A-/E-/VNS | Combined with |
|---|---|---|---|---|---|---|---|---|---|---|---|
| [182] | | | | | | ✓ | | | | | Lower bounding procedure |
| [189] | | | | ✓ | | | | | | | Dynamic Dijkstra |
| [230] | ✓ | | | | | | | | | | ILS |
| [231] | | | | | | | | ✓ | | | |
| [236] | | ✓ | | | | | | | | | |
| [23] | | | | | ✓ | | | | | | |
| [80] | | | | ✓ | | | | | | | LNS |
| [86] | | | | | ✓ | | | | | | Column generation |
| [128, 129] | | | | ✓ | | | | | | | |
| [125] | | | | | | | | | ✓ | | TS |

*ACO* ant colony optimization, *CS* cuckoo search, *DEA* differential evolution algorithm, *GA* genetic algorithm, *ILS* iterated local search, *A-/LNS* adaptive-/large neighborhood search, *MA* memetic algorithm, *SA* simulated annealing, *A-/TS* adaptive-/Tabu search algorithm, *A-/E-/VNS* adaptive-/evolutionary-/variable neighborhood search

total cost determined as a function of speed, gradient, and cargo load. Additionally, Desaulniers et al. [44] tackled EVRPTW by studying various charging policies and applied branch-price-and-cut algorithms to solve them optimally. Hiermann et al. [79] extend a mathematical model of the EVRPTW presented by Schneider et al. [184] by analyzing a heterogeneous fleet of different EVs. They minimize the total distance traveled as well as acquisition costs for each vehicle that leaves the depot. Both studies used ALNS as the solution methodology. Further solution techniques for the EVRPTW concentrating on a math heuristic and a VNS branching are introduced by Bruglieri et al. [24, 25].

Earlier EVRPTW studies by Afroditi et al. [3], Schneider et al. [184], and Bruglieri et al. [24, 25] considered charging at some specific points on the route but neglected recharging at client locations as well as simultaneous partial recharging and siting decisions. Later, the EVRPTW was extended by several authors with different assumptions and formulations. In [164], the authors addressed a load-dependent function for energy consumption in delivery service systems, which was solved by an ATS algorithm. Instead of the battery charging operations of the EVs, Chen et al. [34] studied operations for battery swapping operations. Grandinetti et al. [76] investigated a multi-objective mixed-integer linear programming (MILP) mathematical model for a pickup and delivery problem with soft time windows, through a fleet of EVs. Such a problem tries to minimize the total traveled distance, the vehicle cost, and the penalty cost for violated time windows. Paz et al. [152] considered a multi-depot EVRPTW. Recently, Xiao et al. [216] modeled the EVRPTW considering the maximum range of an EV or the energy/electricity consumption rate (ECR) per unit of distance traveled by an EV dynamically according to speed and load along the route. Keskin et al. [93] investigate time-dependent queueing times at the stations for the EVRPTW with allowing but penalizing late arrivals at the depot and client sites.

Roberti and Wen [174] is the first attempt concentrating on the electric traveling salesman problem with time windows (ETSPTW) which is a special case of the VRP with only a single EV. Since the ETSPTW is known as a relatively new research field, there are just a few studies on the TSP optimization of EVs in the literature. Küçükoglu et al. [125] considered both private and public recharging stations for the ETSPTW and extend it by additionally considering recharging operations at client sites with various recharging rates, called hereafter the ETSPTW with different recharging rates.

## 1.2.1 Refueling Policy

There are three recharging ideas about departure state-of-energy in the tank and the time spent for at a fuel station. In "Fixed-full" refueling approach, which fits BSSs, it has been assumed that a vehicle leaves a fuel station when its tank capacity is filled in a particular constant duration. "Variable-full" approach has a same assumption regarding departure with full state-of-energy but relaxes the constant time assumption. Time spent on refueling under this approach is proportional and depends on the energy required for filling the tank capacity. "Variable-partial" approach allows a vehicle to depart with partial state-of-energy and has a proportional fueling time to energy gain. Below we summarize the main studies on EVs charging policies.

### 1.2.1.1 Full Charging Policy

A full recharging strategy means that an EV is allowed to leave a charging station contingent upon its battery capacity is fully charged. Erdogan and Miller-Hooks denoted the problem as GVRP with a Fixed-full approach, in which the tank capacity of AFV is filled at AFSs within a constant amount of time. Intending to minimize the total distance, the model seeks to eliminate the risk of running out of fuel. They consider the service time of each customer and the maximum duration restriction was posed on each route. Schiffer and Walther [181] analyzed both full and partial charging options for EVs in their location-routing model of charging stations, as well as time window constraints. The objective function aims to minimize not only the traveled distance but also the number of charging stations and the number of used EVs. The researches of Conrad and Figliozzi [39], Schneider et al. [184], Li-Ying and Yuan-Bin [111], Desaulniers et al. [44], Hiermann et al. [79], Koç and Karaoglan [94], Lin et al. [113], Montoya et al. [140], Roberti and Wen [174], Andelmin and Bartolini [7], Breunig et al. [22], Bruglieri et al. [29, 30],

Madankumar and Rajendran [130], Paz et al. [152], Wang et al. [209], Zhang et al. [229], and Basso et al. [14] can be mentioned here.

### 1.2.1.2  Partial Charging Policy

Aside from the full charging assumption, some researches follow partial charging policies where an EV may leave a recharging station with full battery capacity or any level depending on the time spent for recharging. The full charging policy was first relaxed and the EVRP was modeled by allowing EVs to recharge partially by Felipe et al. [53] who applied various recharging technologies for the EVs. The proposed problem aims to minimize the overall costs consisting of costs for the total distance and recharging costs. Various cost terms are supposed for different recharging decisions as the authors assumed that recharging overnight at the depot is more economical than recharging along the route. Besides, many costs and recharging options are surveyed for the different charging technologies including slow and fast charging. Nevertheless, the authors do not take charging at client locations into account and neglect simultaneous siting decisions. To solve the problem, they developed three heuristics including a deterministic local search technique, a construction heuristic, and simulated annealing algorithm. As reported by their computational results, the SA operates more reliably concerning other algorithms for large-scale instances. Moreover, the impact of fast charging options in the presence of time windows has also been investigated by Çatay and Keskin [32].

Further research on partial charging is tackled by Desaulniers et al. [44] concentrating on four different recharging policies per route including only a single full charge, only a single partial charge, multiple full charges, and multiple partial charges. To solve the EVRPTW, they performing an exact method, branch-cut-and-price algorithm. Likewise, Keskin and Çatay [90] investigated the same problem with a partial recharge strategy and developed an ALNS algorithm to solve the problem. The computational results show the better outcomes of the full recharging strategy as well as the benefits of the partial recharging strategy which are pointed out by comparing the partial and full recharging policies. A similar strategy can also be found in [24, 25, 79] that relaxed the full recharge restriction and allowed batteries to be recharged up to any level along the travel route. The former examined also a partial charging strategy in the EVRPTW, where the objective functions minimize the number of used EVs and the time of traveling, waiting, and recharging. To solve small size instances, they developed a VNS branching algorithm. Later, in [25], they model EVRP considering different EVs with various capacities, costs, and recharge rates.

Bruglieri et al. [28] is another research regarding partial recharging strategy where a three-phase metaheuristic approach combined with an exact method and a VNS local branching is presented to solve the proposed EVRP. A different assumption for the EVRP is taken into account by Schiffer and Walther [180, 181] and Schiffer et al. [183] in which siting decisions for the recharge stations are also investigated simultaneously. The first study that extended EVRP to consider nonlinear charging functions is presented by Montoya et al. [141]. The objective function aims to minimize the total time of the operations consisting of charging times and travel times. To solve the problem, the authors developed a hybrid metaheuristic and introduced new benchmark instances. New formulations of this problem are considered by Froger et al. [64, 65]. They developed two new extensions of the EVRP. In the first study, the authors considered nonlinear charging functions, while in the second one, they dealt with an EVRP with capacitated recharging stations.

### 1.2.1.3   Battery Swapping

In the literature, some researchers assumed that the EVs have swappable batteries that can exchange immediately in battery swap stations (BSS), thereby avoiding a long recharging time. Adler and Mirchandani [2] considered a case study of EVs application in an online routing system, where the authors try to optimally route the multiple origin-destination pairs while studying battery swapping and reservation. Zheng et al. [227] introduced an optimal design of battery swap/charging stations for a distribution system. Yang and Sun studied capacitated EVs in a location-routing problem with BSSs, which concentrates on simultaneously determining the routing plan of EVs and the location of BSSs with considering limitations for battery driving range [218]. The EVs always depart from stations with a full battery with considering BSSs instead of charging stations. Besides, in the basic formulation, a BSS can be visited at most once, whereas their proposed problem allowed a BSS to be visited several times.

Another EVRP with battery swapping operations was studied by Liao et al. [108], who tried to find a tour involving the origin and all given points while visiting a set of expected BSSs. To solve the extended model, they presented polynomial-time algorithms that consider the problem with fixed tours where all points are visited in fixed sequences. The EVRP model suggested by Yang and Sun [218] was developed by Schiffer and Walther [181] to consider different effective factors including time window, partial charging policy, and stopping at charging station. Subsequently, Hof et al. [81] investigated a BSS location-routing problem for EVs with intermediate stops and developed an adaptive VNS approach to solving it. Also, Paz et al. [152] modeled the same problem considering multi-depots and battery swapping option in some locations.

Recently, Zhang et al. [232] improved previous deterministic routing models by developing a stochastic routing model applicable to the actual stochastic occurrences. They present an electric vehicle BSS location-routing problem with stochastic demands, intending to determine a minimum cost scheme including the optimal number and location of BSSs with an optimal route plan based on stochastic customer demands. Furthermore, the classical recourse policy and preventive restocking policy are extended by considering the influences of both battery and vehicle capacity simultaneously. To solve this problem, a hybrid approach was proposed that combines the binary PSO algorithm and VNS algorithm.

## 1.2.2 Energy Consumption Rate

Because of the limited battery capacity and the resulting shorter achievable range of EVs, such routing problems require new approaches that can accurately monitor the vehicles' energy consumption to ensure they never get stranded. For example, Barco et al. [10] proposed an EVRP, which minimizes the used energy instead of the traveled distance. An approximated energy consumption model based on physical dependencies is used to compute energy consumption. Since the model is applied to an airport shuttle service that carries passengers from the airport to a hotel and back at known times, recharging is only possible at the depot. They conduct a scheduling model for this application. In this model, the recharging of vehicles on known routes is scheduled, taking battery degradation effects into account. Due to the application, the approach does not regard the usage of charging stations on routes or any siting decisions.

Some studies modeled the EVRP by a linear battery discharges rate concerning traveled distance. For example, Felipe et al. [53] assumed various technologies for battery charging where each of them is related to a constant charging rate and has a fixed unit cost. Others have considered a more detailed energy consumption model based on the comprehensive emissions model of Barth et al. [13] which computes the required energy to traverse any given arc based on certain vehicle and arc characteristics (e.g., mass, speed, elevation, acceleration, rolling friction, frontal area, air drag). This energy consumption model was first used in a VRP setting for the pollution-routing problem to estimate fuel consumption and emissions [17], and subsequently in many EVRP studies to track the state-of-energy consumption [14, 71, 100, 164].

In all of the discussed studies, there is a simple assumption about the energy consumed by a vehicle over a route which is assumed to be a linear function of the traveled distance. The earliest research on the modeling of a more realistic energy consumption approach for the EVRP was presented by Goeke and Schneider [71] who designed a nonlinear ECR for a mixed fleet of EVs and fuel vehicles. Although they considered practical factors such as the gradient of the terrain, the vehicle speed and load, the travel speeds/times are considered constants and previously known. Therefore, only the weight of the cargo affects energy consumption over a route. To determine the energy consumption of EVs, Fetene et al. [57] harnessed big data and showed that the ECR may change considerably with a nonlinear behavior depending on the driving patterns and weather conditions.

Lin et al. [113] is another study that developed energy consumption model for EVs, which considered the effects of vehicle load on battery consumption. Fiori et al. [59] introduced a comprehensive power-based electricity consumption formulation that can estimate the instantaneous ECRs by evaluating inputs of EVs (vehicle mass as well as vehicle speed and the instantaneous acceleration). A physical-based analytical approach for calculating energy consumption was developed by Yuan et al. [226]. The authors traced ECRs based on the energy flow path of EV during a driving cycle which includes driveline losses, electric motor losses, battery losses, load losses, and brake losses. Liu et al. [110] studied the effect of the road gradient on the energy consumption of an EV. Pelletier et al. [156] integrated battery degradation into the model and optimized charging schedules at the depot. They also provided managerial insights considering degradation, grid restrictions, charging costs, and charging schedules of the fleet.

To the best of our knowledge, only Fontana [60] has considered the uncertainty surrounding the energy consumption of electric vehicles and has modeled the problem within a robust optimization framework. However, this author considers the idea of uncertainty surrounding the energy consumption of EVs in an optimal path problem setting, while Pelletier et al. [157] extended it to an EVRP setting in which the vehicle loads influence energy consumption.

## 1.2.3 Fleet Size and Mix

As EVs have gained popularity worldwide over the past decade, the issues related to the usage of EVs for transportation have opened up a wide range of relevant research works. In addition to the mentioned strategy for EVs, some of the researchers pay attention to the impacts of fleet characters on routing plans and incurred travel costs. Contrary to Van Duin et al. [205] which did not consider battery charging, assumed charging could occur anywhere [73] or only at the depot between routes [100], other fleet size and mix studies incorporating EVs have considered the possibility for the vehicles to stop at charging stations during their routes and have taken into account the actual location of the stations.

Pelletier et al. [154] provided researchers with the technological and marketing background of goods distribution with a fleet of EVs and presented a review of existing research works in the EVRP field. Their overview summarized the main existing research works on EVs in transportation science, which covers three parts: fleet size and mix, vehicle routing, and optimal paths. From their work, they found that EVs have many advantages such as frequent stop-and-go movements, low travel speeds and reduced pollution, and noise in last-mile deliveries of urban logistics.

A different variant of the problem with a mixed fleet of EVs and conventional ICEVs was developed by Goeke and Schneider [71] who used realistic energy consumption functions for both types of vehicles, considering vehicle mass as well as vehicle speed and terrain gradients. Three different objective functions are presented. The first objective minimizes the total traveled distance, whereas the second objective minimizes overall costs as the sum of energy costs and wages for drivers. The third objective extends the second one by adding costs for battery lifetime. However, options of partial recharging and simultaneous siting decisions are not taken into consideration.

Lebeau et al. [100] have studied the fleet size and mix EVRPTW. The fleet is composed of EVs and ICEVs each having their fixed costs, operating costs, payload, volume capacity, energy capacity, and energy consumption. The objective is to minimize the sum of all fixed, operating, and driver costs. Each customer has a demand, a time window, and a service time. The depot also has a service time representing loading time so that vehicles can accomplish several routes. The EVs can be charged at the depot during loading with a linear charging rate. They solved the problem using a modified Clarke and Wright Savings (MCWS) algorithm.

Another study that considers different types of EVs in a heterogeneous fleet for customer visits was done by Küçükoglu and Öztürk [124]. Compared with the computational results on small-scale sample sets generated via the mathematical formulation suggested by Schneider et al. [184], the authors pointed out the benefits of using the heterogeneous fleet on the number of used EVs and the total traveled distance. To solve the same problem with a heterogeneous fleet, Penna et al. [158] proposed a hybrid iterative local search (ILS) algorithm developed by combining an ILS algorithm with a set partitioning problem, and Hiermann et al. [79] offered an efficient ALNS algorithm whose performance has been demonstrated in several sample sets.

## 1.3   Green Vehicle Routing Problem

As the energy overuse and generated pollution have a potential threat for our environmental and ecological condition, many researchers have taken the initiative way to join the green campaign to prevent more damage to the environment. Although barriers to achieving sustainable solutions still exist, energy policies need to be changed to reduce fossil fuel consumption and encourage the use of AFVs. In the last decade, regarding the growing knowledge about the hazardous effects of logistics activities, such as transportation which has a considerable impact on the ecological environment, requests for more environmentally friendly practices have increased [112]. In this way, Erdogan and Miller-Hooks [48] introduced the GVRP, a relatively new research field, as an extension of the traditional routing problem. To decrease energy consumption in fleet operations, GVRPs aim to minimize fuel consumption or adopt AFVs solely or with existing conventional gasoline or diesel vehicles (GDVs) in fleets.

This section illustrates transportation and energy consumption, followed by a comprehensive survey on the recent investigations on the pollution-routing problem. For earlier surveys, interested readers can refer to [112]. Figure 1.2 displays the trend of scientific contributions (research papers, review and book chapters) on GVRP published in Scopus-indexed journals. As shown in the figure, over the years, the green concept and its several variants have received increasing attention in the scientific context, especially in the field of operations research [18, 43, 112, 151].

Table 1.5 reviews the relevant works in terms of the characteristics of the mathematical model, fleet type, fueling approach, and solution method. Definitions of the numbers associated with each objective function have been presented in Table 1.3.

As shown in Table 1.5, most GVRPs examined vehicle weight, speed, and travel time as effective factors on greenhouse emissions [18, 112, 151, 225]. Various

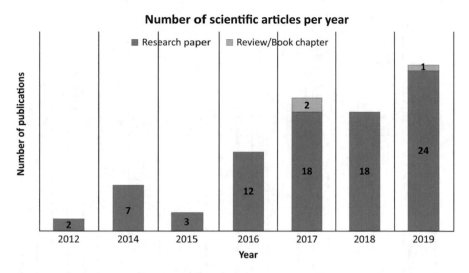

**Fig. 1.2** Studies in the context of GVRP and its variants

VRP characteristics have been incorporated into the GVRP literature. These include customer time windows and vehicle capacities [27, 71, 79, 90, 184, 185], multiple charging technologies [53], speed- and/or load-dependent energy consumption [71, 162], nonlinear recharging times [141], recharging station congestion [45, 64, 65], time- and/or station-dependent refueling costs [56, 91].

Bektas et al. [18] studied GVRP where $CO_2$ emissions are explicitly taken into account. They reviewed some recent developments in the field, including the description of some vehicle-emission models and their applications in road freight transportation. To minimize the carbon emissions for a green pickup and delivery problem with the lowest carbon emissions, Yu et al. [222] applied integrated scheduling. Qazvini et al. [166] proposed a multi-depot green pickup and delivery problem along with time windows to lessen fuel consumption while a GVRP with cross-docking was investigated in [221]. Sawik et al. [178] concentrated on multi-objective GVRP optimization consisting of the minimization of pollution, noise, and fuel consumption. Wang et al. [209] addressed the compensation and profit distribution by using the cooperative game theory for the cooperative green pickup and delivery problem with the lowest carbon emissions.

In comparison with other variants of GVRPs, the literature that considers halts for refueling/recharging at fuel/charging stations is limited. Erdogan and Miller-Hooks [48] examined the possibility of recharging or refueling a vehicle with environment-friendly fuel like biogas as well as the halts for refueling at stations. They conceived the main idea from Bard et al. [12] who addressed VRP with considering maximum carrying load capacity for the fleet.

Lin et al. [112] considered charging stations in the tour to overcome the capacity limitation of the fuel tank. The approach of Chung and Kwon [36] considers the

**Table 1.5** Summary of the previous work on the GVRP

| Study | Fleet type Ho | Ht | Capacity | Load | Speed | Time window | Multi-obj | Objective(s) | Solution method |
|---|---|---|---|---|---|---|---|---|---|
| [63] | AFV | | ✓ | ✓ | | ✓ | | 3 | An exact algorithm |
| [122, 123] | GDV | | ✓ | ✓ | ✓ | ✓ | ✓ | 1, 3 | ILS based metaheuristic |
| [185] | AFV | | ✓ | | | ✓ | | 2 | AVNS |
| [69] | GDV | | ✓ | ✓ | ✓ | ✓ | | 1, 3 | Exact method |
| [195] | GDV | | ✓ | ✓ | ✓ | | ✓ | 3 | SA & TS |
| [37] | GDV | | ✓ | ✓ | ✓ | | ✓ | 3, 8 | Dynamic programming |
| [49] | GDV | | ✓ | ✓ | ✓ | ✓ | | 3 | Exact method |
| [62] | GDV | | ✓ | ✓ | ✓ | ✓ | ✓ | 1, 3 | ALNS |
| [84] | GDV | | ✓ | ✓ | | | ✓ | 2, 3 | ACO & VNS |
| [191] | GDV | | ✓ | ✓ | ✓ | | | 2 | Dynamic programming |
| [201] | GDV | | ✓ | ✓ | ✓ | | ✓ | 3, 12 | Exact method |
| [219] | | AFV & GDV | | | | | | 2 | Heuristic procedure |
| [220] | | AFV & GDV | ✓ | ✓ | | | | 2, 3, 8 | VNS |
| [119] | GDV | | ✓ | ✓ | ✓ | ✓ | ✓ | 2, 3, 12 | NRGA, NSGA-Π & PSO |
| [138] | GDV | GDV | ✓ | ✓ | ✓ | ✓ | | 2 | Exact method |
| [147] | GDV | | ✓ | ✓ | ✓ | ✓ | | 1, 3 | TS |
| [162] | GDV | | | ✓ | ✓ | | ✓ | 2, 3 | PSO |
| [229] | GDV | | ✓ | ✓ | ✓ | ✓ | | 3, 8, 11 | ALNS |
| [47] | GDV | | ✓ | ✓ | ✓ | ✓ | | 3, 12 | Local search |
| [106] | AFV | | ✓ | ✓ | | | ✓ | 2, 3, 7, 11 | ACO |
| [121] | | AFV & GDV | ✓ | | | ✓ | | 2 | Branch-and-cut |
| [170] | | GDV | | ✓ | ✓ | ✓ | | 1, 3, 7 | Exact method |
| [172] | GDV | | | ✓ | ✓ | | | 2, 3 | NSGA-Π |
| [210] | | AFV & GDV | | ✓ | | | | 3 | TS |
| [233] | AFV | | | | | | | 2 | ACO |

fact that the installation of a network of stations is a long-term process that usually requires several periods to be completed. Hence, their model restricts the number of newly established stations per period while pursuing maximization of possible traffic over all periods. Bruglieri et al. [26] contributed a new formulation of GVRP which includes a precomputation of AFS to locate between each pair of vertices. Their model avoids using multiple copies of AFS that leads to a decrease in the number of decision variables. Recently, Poonthalir and Nadarajan [163] intended to address the issues that are associated with the waiting time at the fueling station. Each fueling station was modeled as M/M/1 queue model, where the vehicle is allowed to wait in the queue.

In GVRP, alternative-fueling stations are copied multiple times and treated as special customers (dummy nodes) which are visited at most one time. Therefore, the GVRP was converted to be under the traditional VRP frameworks and it can be solved by employing efficient heuristics proposed for VRPs. Schneider et al. [184] extended GVRP with considering time window constraint and recharging stations. They developed TS with a variable neighborhood algorithm to solve the problem and got better results than density-based clustering and MCWS algorithms presented by Erdogan and Miller-Hooks [48]. Later, Schneider et al. [185] developed VRP with intermediate stops and solved the problem using an adaptive variable neighborhood search algorithm. Montoya et al. [140] developed a simple but effective two-phase heuristic to solve the GVRP. They also performed extensive experiments on 52 instances from the literature to test their heuristic.

Frank et al. [63] introduce an arc-duplicating formulation for a rich VRP with time windows, considering capacity restrictions including cargo and energy, customer time windows, and possible recharging stops. Koç and Karaoglan [94] adapt the arc-duplicating formulation of Frank et al. [63] to the GVRP. They used a SA based exact solution approach using branch and cut technique to solve the proposed problem. Yavuz [219] developed an iterated beam search algorithm for the GVRP based on arc duplication.

In the literature, there are some papers that have used set partitioning to solve GVRP types. Montoya et al. [140] proposed a multi-space sampling heuristic for GVRP. In particular, a randomized route-first cluster-second heuristic is firstly used for generating a set of feasible single-vehicle routes. Then, the feasible set is fed into a set partitioning formulation for determining a solution of the GVRP. The model focused on minimizing the length of the route and carbon dioxide-equivalents emission. They showed that it outperforms all other heuristics and meta-heuristics designed for the GVRP. Very recently, in [80], an extended fleet for including conventional, PHEVs, and EVs was proposed which was heuristically solved via a hybrid GA enhanced with local and LNS and set partitioning. Exact solution via branch-and-price-and-cut has also been developed.

Desaulniers et al. [44] tackled four problem variants with time windows, which are the combinations of single and multiple visits of a recharging station and partial and full charge allowances, while Andelmin and Bartolini [7] tackled the basic form of GVRP. A key difference between two proposed exact approaches for solving these problems is that Desaulniers et al. [44] used node duplication in their pricing problem, whereas Andelmin and Bartolini used arc duplication in a way that includes multiple external stations visits back-to-back under the fixed-full refueling approach. They formulated a set partitioning problem where columns illustrated the feasible routes like simple circuits in multiple graphs in which each node is associated with a customer while an arc represents a non-dominated path between two customers that possibly uses AFSs. The set partitioning formulation is strengthened through several valid inequalities and it is then solved with a CG method. Their computational experiments showed that the proposed exact solution approach is suitable to optimally solve some large-scale instance sets in an average computational time of 3 h. As reported in [115], CG approaches are very effective but their convergence is very slow. Thus, they are generally suitable to solve large-sized instances in long computational times, but they are not suited to solve smaller-sized instances in very fast computational times. Fukasawa et al. [68] proposed a branch-and-cut-and-price algorithm for the energy minimizing vehicle routing problem (EMVRP) in which the energy consumption over each arc is defined as the product of the arc length and the weight of the vehicle. They developed two new MILP formulations for the problem: an arc-load formulation and a set partitioning formulation.

A path-based MILP formulation was done for GVRP by Pisacane et al. [159]. They generated all feasible routes and eliminated all dominated paths from the feasible set which is given as an input to a set partitioning formulation that gives solutions to the problem. Leggieri and Haouari [101] proposed a new MILP of GVRP for considering time and energy consumption constraints that is neither node- nor arc-duplicating. Their formulation reduced the use of various variables and constraints and included a set of preprocessing conditions which made the problem to be efficiently solvable using commercial solvers.

Another problem of GVRP dealing with the recharging or refueling of the vehicles, particularly the AFV, is charging infrastructure planning (e.g., [75]). Government agencies, nonprofit organizations, municipalities, and some private companies have started to convert their fleets of trucks to AFVs so as either to satisfy the energy policies or environmental regulations or to voluntarily reduce the environmental impact [48]. GVRP is mainly focused on alleviating the negative influences on the environment associated with conventional vehicles. One way of alleviating such impacts is to employ the AFVs. Compared with conventional vehicles, the AFVs use clean energy like electricity and hydrogen as power sources. However, the AFVs have some shortcomings such as limited energy storage (e.g., low battery capacity of electric vehicles) and access to AFSs. Consequently, the refueling issue becomes more important in solving the routing problem of the AFVs, which makes the route selection for the AFVs complicated. The problems with capacitated AFSs have been introduced also for the GVRP by Bruglieri et al. [27]. In [130], two MILPs have been proposed for routing a set of AFVs considering pickup and delivery operations in a semiconductor supply chain. In Soysal and Çimen [37], a green time-dependent capacitated vehicle routing problem (CVRP) has been addressed through dynamic programming.

The time-dependent GVRP literature has mainly focused on time-dependent travel times due to congestion on roads. The studies can be classified as static or dynamic if the times are fixed for a given time interval, or change dynamically, respectively. Static mathematical models include deterministic [62, 83, 93, 162, 170, 192, 211] or stochastic features [37, 169] whereas [191] dealt with dynamic models. A periodic GVRP with time-dependent urban traffic and time window was studied by Mirmohammadi et al. [139]. They modeled the problem to minimize carbon emission, earliness, and lateness penalties and demonstrated their results for some generated test problem instances.

## 1.3.1   Transportation and Energy Consumption

Transportation, which is one of the most important parts of logistics, is the irreplaceable fundamental infrastructure for economic growth. However, it is also one of the hugest petroleum consumers and accounts for a large part of the overall pollutants [175]. Some green routing problems focus on energy consumption. Reducing fuel consumption may clash with the designated economic objectives. Exploring the relationship between environmental effect and transportation through route planning will be able to provide practical and valuable suggestions. It is also desirable that a decrease in petroleum-based fuel consumption can correspondingly reduce the greenhouse gas emission significantly [215]. Therefore, fuel consumption is an important index in the GVRP [126]. To include the fuel consumption in the routing model, the formulation of computing fuel consumption concerning the condition of a traveling vehicle is essential. According to the report by the US Department of Energy (2008), the weight of the load, travel speed as well as the traveled distance are influential parameters and the key factors that considerably affect the amount of the fuel consumption.

Some studies about the fuel consumption model in terms of transportation exist in the literature, which provides relevant references to the research on GVRP. Just to cite a few, a formulation of fuel consumption is provided in [215]. They proposed an ECR that extends CVRP with the aim of minimizing fuel consumption. In the paper, both the load and the distance traveled are considered as the factors which determine the fuel costs. Fuel consumption rate is taken as a load-dependent function, where the consumption rate is linearly associated with the vehicle's load. Tiwari and Chang [200] considered a GVRP model with the emissions minimizing objective and estimated the carbon dioxide emissions based on truckload and traveling distance. A block recombination approach is used to solve the proposed problem. In addition to the loading weight and the transportation distance that are considered in the above two papers, Kuo [126] added the transportation speed to the fuel consumption calculation model in time-dependent VRPs. Other VRP-related researches that aim at minimizing total fuel consumption are [8, 50, 135, 145, 176, 198]. Demir et al. [41–43], and Bektas and Laporte [17] applied a comprehensive mathematical model for estimating fuel consumption as an important parameter for $CO_2$ emission. Scott et al. [187] and Jovicic et al. [88] developed a computer application to import data from the road and measure directly fuel consumption. Also, Jabali et al. [83] and Pan et al. [150] proposed a new approach to calculate the fuel consumption which is proportional to the amount of $CO_2$ emission. Some survey papers that focus on different fuel consumption formulations are [18, 43]. To decrease the amount of fuel consumption in a time-dependent problem, Kuo [126] applied a string-based SA on retail stores. Their formulating is based on the calculation done by Kuo [126]. Suzuki [194] investigated a multi-stop truck routing problem consisting of time-constrained transportation which aims to reduce fuel consumption. Xiao et al. [215] presented a time-dependent mathematical model that finds the best routes with the lowest fuel consumption and pollutant emission. To calculate fuel consumption, the authors employed a regression formulation which had an improvement of 24.61% in fuel consumption with better routing plans but longer transportation time and distance. Since this consumption model does not use the deceleration, recently Suzuki and Lan [196] extended it, by including negative values of acceleration to study the fuel consumption of trucks in congested areas. They ensured that the fuel use rate cannot be lower than that of engine idling, to avoid the unreal case of negative fuel burn during the deceleration phase.

Kara et al. [117] examined developing an EMVRP which focused on minimizing the weight of loads that leads to decreasing fuel consumption. To solve a model for minimizing fuel consumption in a routing problem with a time window, Li [102] developed a TS algorithm with random variable neighborhood decent procedure. Teng and Zhang [199] proposed a GVRP with a load factor that tries to minimize driving distance and total fuel cost and solved using SA. Jabir et al. [84] developed a hybrid ACO for capacitated multi-depot GVRP to decrease emission cost. Kazemian and Aref [120] solved a capacitated time-dependent mathematical model that aims to minimize both fuel consumption and fuel emission.

## 1.3.2 The Pollution-Routing Problem

The growing concerns about direct or indirect hazardous effects of transportation on humans and the whole ecosystem require re-planning of the road transport network and flow by explicitly considering Greenhouse Gas (GHG) and in particular $CO_2$ emissions [17]. The Pollution-Routing Problem (PRP) aims at choosing a vehicle dispatching scheme with less pollution, in particular, reduction of carbon emissions. Reducing $CO_2$ emissions is achievable by extending the traditional routing problem to take into account relevant environmental and social impacts [40, 69, 131, 137, 149, 167, 168, 179]. Although the traditional routing problem objective of minimizing the total distance will in itself contribute to a decrease in fuel consumption and pollutant emissions, this relationship needs to be directly calculated using more accurate formulations.

Pronello and André [165] suggested that reliable models for estimating the pollution generated by vehicle routes need to consider traveling time when the engine is cold. Sbihi and Eglese [179] considered a time-dependent routing problem in the context of traffic congestion. Since less pollution is produced when the vehicles are at best speeds, directing them away from congestion tends to be more environmentally friendly, even though it leads to longer traveling distance. Also, Maden et al. [131] solved the same problem and reported about 7% reduction of $CO_2$ emissions based on a function conducted from the time-varying speeds. However, the objective function of their model remained the minimization of the total travel time instead of reducing emissions. Palmer [149] developed a routing model and calculated the amount of carbon dioxide as well as distance and traveling time. The paper examined how speed affects the reduction of $CO_2$ emissions in different congestion scenarios with time windows.

A fuel emission minimization model that concentrated on reducing emission for solid waste collection trucks was done by Apaydin and Gonullu [8]. An emission minimization problem was proposed by Figliozzi [58] that examined routing vehicles with less emission where emission minimization is taken as an additional objective. The results illustrated about 5% reduction of the $CO_2$ emissions emitted on the journey. Bauer et al. [15] explicitly concentrated on minimizing GHG emissions in a model of intermodal freight transport, showing the potential of intermodal freight transport for reducing greenhouse emissions. Fagerholt et al. [51] tried to reduce fuel consumption and fuel emissions by optimizing the speed in a shipping scenario. Given the fixed shipping routes and the time windows, the speed of each segment of a route is optimized to yield fuel savings.

Ubeda et al. [202] is the first study that incorporates minimizing GHG emissions in the model of routing problem with backhauls. They conducted a comprehensive objective function that measures both the distances and pollutant emissions. The results also revealed that backhauling seems more effective in controlling emissions. The authors suggest that backhauling can be initiated by companies to enhance energy consumption efficiency and reduce environmental impact. Faulin et al. [52] developed a CVRP with environmental criteria and considered a more complex environmental impact. Apart from the traditional economic cost's measurement and the environmental costs that are caused by polluting emissions, the environmental costs derived from noise, congestion, and wear and tear on infrastructure were also considered.

Following up Bektas and Laporte [17] and Demir et al. [41] proposed an extended ALNS for PRP to minimize fuel consumption, emission, and driver cost and enhance the computation efficiency for medium or large-scale PRP. The algorithm for solving PRP iterates between solving VRPTW and speed optimization. Initial routes are formed for VRPTW using fixed speed and are improved using a speed optimization algorithm. They further extended the work to solve PRP as a bi-objective optimization problem [42] that concentrated on arriving at a tradeoff between the conflicting objectives of fuel consumption and total driving time. The problem is modeled with four different a posteriori methods using an enhanced ALNS procedure where speed optimization is done at each iteration. They have used a comprehensive emission model to estimate fuel consumption.

A bi-objective GVRP using NSGA-II was formulated by Jemai et al. [85] with objectives to minimize distance and $CO_2$ emission. A green vehicle distribution model for the public transport network was developed as a nonlinear optimization problem by Jovanovic et al. [87]. They designed an Adaptive Neuro-Fuzzy Inference System (ANFIS) model with environmental parameters, passenger costs, and their impact on green routing to solve the problem. Cirovic et al. [38] concentrated on routing light delivery vehicles using an adaptive neural network that focused on reducing air pollution, noise level, and logistics operating costs. They pondered routing limited environmentally friendly vehicles and unfriendly vehicles separately. The input parameters for the neural network were logistics operating costs and environmental parameters which were used to assess the performance of network links for calculating routes and then a MCWS algorithm was used. A Satisfactory-GVRP was developed by Afshar-Bakeshloo et al. [4] as an extension of PRP that takes into account the economic, environmental, and customer satisfaction as objectives.

Fukasawa et al. [69] developed two mixed-integer convex optimization models for the PRP. Dabia et al. [40] presented a branch-and-price algorithm for the PRP, and a new dominance criterion for the tailored labeling algorithm was developed to solve the pricing problem. Also, Majidi et al. [133] presented a nonlinear mixed-integer programming model for PRP. In another study, Franceschetti et al. [62] considered traffic congestion and proposed an enhanced ALNS algorithm for the time-dependent version of the PRP, which is an extension of [61]. Considering the demand and travel time uncertainty in green transport planning, Eshtehadi et al. [49] proposed several robust optimization techniques to solve PRP. Kargari Esfand Abad et al. [119] proposed a bi-objective model for pickup and delivery PRP to minimize the total system cost and fuel consumption to reduce the carbon emission. Van den Hove et al. [204] described the development of a cyclist routing procedure that minimizes personal exposure to black carbon. They indicated that the average exposure to black carbon and the exposure to local peak concentrations on a route are competing objectives, and proposed a parametrized cost function for the routing problem that allows for a gradual transition from routes that minimize average exposure to routes that minimize peak exposure.

All of the above studies are based on the assumption that there is a homogeneous fleet of vehicles. The PRP is one important variant of the GVRP that specifically necessitates a heterogeneous fleet of vehicles; firstly, because vehicles may produce a variety of emissions; and secondly, because logistic agencies may acquire heterogeneous vehicle types involving various categories of engines (fuel, electric, or hybrid), recent and old models, as well as vehicles from different brands [89, 95, 98, 127]. Pitera et al. [160] formulated an emission minimization routing decision for urban pickup system with a heterogeneous fleet. Their study demonstrated a significant improvement in emission reduction. Xiao and Konak [214] presented a time-dependent heterogeneous green vehicle and scheduling problem, with the objectives to minimize $CO_2$ emission and weighted tardiness. Travel schedules of vehicles were determined using traveled distance in different periods and solved using a dynamic programming approach. Yu et al. [224] propose an improved branch-and-price algorithm to precisely solve the heterogeneous fleet green vehicle routing problem with time windows. The latest studies about the heterogeneous GVRP, where $CO_2$ emissions reduction are considered, can be found in the work of Yu et al. [225] and Li et al. [105].

The alternative fuel in routing problems is also studied under the category of the PRP. Although, the models in the PRP commonly consider a constant ECR and assume a linear relationship between energy consumption and the traveling distance of vehicles. Bektas and Laporte [17] proposed the PRP that considers both speed and load as main factors in emission and introduced a comprehensive mixed-integer nonlinear programming model. The objective function presented the minimization of the cost of carbon emissions along with the operational costs of drivers and fuel consumption. However, their model assumed a free-flow speed of at least 40 km/h, which was contrary to the real-world situation where congestion occurs. They adopted a fuel consumption function that is inspired by the comprehensive modal emission model (CMEM) developed by Barth et al. [13]. The main advantage of using the CMEM is twofold; first, the CMEM depends on the fuel type; second, biodiesel is viewed as an alternative fuel that can be used in conventional diesel engines, either on its own or mixed with diesel [206]. Besides, the properties and suitability of biodiesel do not require any adjustments in the engine of a specific vehicle due to the use of biodiesel instead of (or blended with) petroleum-based diesel. Thus, the new diesel-powered vehicles are designed to operate on biodiesel without modification [16]. For additional information on biodiesel fuels, the readers are referred to [132, 217].

## 1.4  Hybrid Vehicle Routing Problem

When a vehicle equipped with artificial fuel needs to take a long-distance tour, it has to halt for refueling in an alternate fueling station. When AFS is limited, a vehicle has to take up a route forcefully to refuel in the limited number of AFS which can increase the route cost. This problem can be tackled if the vehicle can switch between battery and fuel depending on the requirements, called the Hybrid Electric Vehicle (HEV), as proposed by Mancini [134]. In [134], a MILP formulation of the HVRP is carried out where the decision process is related to deciding both when to recharge and when to switch from an ICEV to an EV. The author proposes a large neighborhood search based metaheuristic and she also uses the GVRP benchmark instances for the experimental campaign.

Although, a HEV is preferable because a pure electric vehicle is limited to the availability of recharging stations and distance coverage, the availability of a routing model for HEVs is even less or quite absent. Therefore, this section is motivated to review the HVRP which is classified into the GVRP group by Lin et al. [112] as it also takes into account the environmental impact. HVRP aims to minimize the total travel cost incurred by utilizing hybrid vehicles. The practicality of HVRP is supported by the growing number of HEVs in the automobile market. For example, Juan et al. [89] introduced a VRP with multiple driving ranges, which was motivated by the practical challenge driven by the increasing number of EVs with different degrees of autonomy in corporate fleets. They described an integer programming formulation and a multi-round heuristic algorithm that iteratively constructed a solution for the problem. Using a set of benchmarks adapted from the literature, the algorithm was then employed to analyze how distance-based costs are increased when considering "greener" fleet configurations.

Lin et al. [234] combined the practical application of HEVs in the logistics industry and studied the HVRP considering battery capacity constraints, gasoline constraints, and mode selection in HEVs based on the background of green logistics and the characteristics of HEVs, which minimized total energy consumption cost. The mode selection system can adjust the driving mode of the HEV according to different road conditions to obtain the optimal use of fuel. This problem was formulated as a mixed-integer linear programming model. To solve the problem, an improved PSO algorithm is developed in which a labeling procedure is involved.

While limited work is devoted to studying the HVRP, several versions of the classical VRP share features of the HVRP. For example, the GVRP presented by Erdogan and Miller-Hooks [48], Schneider et al. [184], and Hiermann et al. [79] entails scheduling efficient routes for EVs that are required to stop at charging stations distributed in the network to recharge their batteries. Failing to schedule proper stops for battery recharging precludes the vehicles from completing their scheduled tour and/or returning to their depots. As such, scheduling stops for battery recharging is a hard constraint for the vehicles in the GVRP. A similar constraint should be considered for the vehicles in the HVRP as their batteries need to be recharged to complete their tours. Besides, the HVRP is similar to the CVRP, where each vehicle has a limited carrying capacity [19, 35, 67, 215]. All studies on HVRP are shown in Table 1.6.

Several sources of complexity characterize the HVRP. First, even for small size problems, the HVDRP involves a large number of decision variables including vehicle resources/capabilities, locations of vehicle dispatching and collection, and routing decisions for the vehicles. The problem can be generally viewed as an extension of the classic VRP which is known to be an NP-hard problem [72]. Thus, the execution time required to obtain an exact optimal solution grows exponentially

**Table 1.6** The studies on the HVRP and its variants

| Year | Number of papers | The list of the papers |
| --- | --- | --- |
| 2007 | 2 | [33, 186] |
| 2008 | 1 | [74] |
| 2009 | 2 | [20, 190] |
| 2010 | 1 | [6] |
| 2013 | 1 | [143] |
| 2014 | 2 | [77, 89] |
| 2015 | 1 | [109] |
| 2016 | 1 | [146, 193] |
| 2017 | 2 | [134, 223] |
| 2019 | 4 | [106, 107, 118, 234] |

as the problem size increases. Second, most decision variables involved in this problem are highly interdependent and cannot be optimized separately.

A hybrid vehicle routing model was set up by Liu et al. [109] to solve the hot strip mill scheduling problem which was a combination of variable fleet VRP, prize collection vehicle routing, and CVRP. The constraints in hot mill processes were analyzed in three aspects: process scheduling, product means, and energy consumption. They proposed a new algorithm combining PSO and SA to solve a multi-objective optimization problem. The PSO solution performance was an imbalance phenomenon between the convergence rate and the convergence precision, due to the inherent mathematical operators of the PSO algorithm. Recently, Li et al. [107] minimized total travel costs in consideration of both electric charging stations and fuel stations. To solve the problem, they proposed a novel hybrid metaheuristic approach which was a hybridization of the memetic algorithm with sequential variable neighborhood descent.

There are two basic configurations of HEV: a series hybrid and a parallel hybrid. In a series hybrid, the power transmission comes from both combustion engines, and the drive system is from electricity. In a parallel hybrid, the power transmission comes from both combustion engines, and the electrical drive system is mechanical. These systems can be applied in combination or independently. Another hybrid configuration is a complex configuration, which involves a setting that cannot be classified into those two types mentioned. The complex hybrid seems to be similar to the series–parallel hybrid because the generator and electric motor are both electrical and mechanical. However, the key differences are the bidirectional power flow of the electric motor in the complex hybrid and the unidirectional power flow of the generator in the series–parallel hybrid [33, 77] classified HEV, based on the hybridization factor, into Micro-HEVs, Mild-HEVs, Power-assisted HEVs, and Plug-In HEVs (PHEVs).

A PHEV is the next generation of hybrid electric vehicles, offering significant advantages over the cleanest and most efficient of today's hybrid vehicles [6]. PHEV's sources of energy for propulsion consist of electricity from the grid, energy recovery while braking (which is stored in the battery), and chemical energy from a hydrocarbon-based fuel stored in the fuel tank [190]. Bradley and Frank [20] explained that a PHEV can work in several modes, such as charge-depleting mode, charge-sustaining mode, EV mode, and engine-only mode. Silva et al. [190] classified PHEV into a range extender PHEV and a blended PHEV. A range extender PHEV runs primarily on electricity discharged from the battery, and after the SOC reaches a minimum, it operates in a charge-sustaining mode. A blended PHEV is similar to the range extender PHEV, but in the charge-depleting mode, it may turn on the internal combustion engine whenever needed.

Yu et al. [223] focused on vehicles that use a hybrid power source, known as the PHEV and generate a mathematical model to minimize the total cost of travel by driving PHEV. They considered a time constraint and two types of energy resources: electricity and fuel. To solve the problem, they developed SA with a restart strategy.

The proposed SA algorithm was first verified with benchmark data of the CVRP, with the result showing that it performs well and confirms its efficiency in solving CVRP.

The routing algorithms for PHEVs should also account for the significant energy efficiency differences of different operating modes and recommend the predominant mode of operation for each road segment during route planning. The energy efficiency differences are also a function of the vehicle (e.g., payload) and road segment features (e.g., speed limits, terrain geometry). Given that internal combustion engines tend to be most efficient when operating at steady highway speeds of 45–65 mph (miles per hour), the electric mode is relatively efficient on city roads with lower speed limits [161].

The strategy to control the energy among these multiple energy sources of a hybrid vehicle is termed "powertrain energy management." An overview of this area is provided by Sciarretta and Guzzella [186]. Currently, powertrain energy management control algorithms for PHEVs are mostly "static" and try to utilize the charge within the battery as soon as possible, without explicit consideration for real opportunities that might be present within the route to best utilize the battery charge. This is also attributable to the fact that there exist no route planning algorithms for PHEVs, and automotive original equipment manufacturers have yet to develop "dynamic" energy management control algorithms that account for the complete route plan in controlling the vehicle powertrain operating modes. Several researchers investigated the powertrain management for hybrids and PHEVs from an energy management control perspective (e.g., [74, 143]). Nejad et al. [146] addressed the energy-efficient routing of PHEVs during route planning, which is a higher-level energy management process for PHEVs and designed exact and approximation routing algorithms that consider general graphs.

The existing shortest-path based algorithms cannot be applied to solving problems for PHEVs, backup batteries with combustion engine, because of the several new challenges: (1) an optimal route may contain circles caused by detour for recharging; (2) emissions of PHEVs not only depend on the driving distance, but also depend on the terrain and the SOC of batteries; (3) batteries can harvest energy by regenerative braking, which makes some road segments have negative energy consumption. To address these challenges, Sun and Zhou [193] proposed a green navigation algorithm (GNA) that found the optimal strategies: where to go and where to recharge. GNA discretized the SOC, then made the PHEV routing problem to satisfy the principle of optimality. Finally, GNA adopted dynamic programming to solve the problem.

## 1.5 Conclusions

It is evident that the managing of fuel consumption and the use of sustainable energy sources in road logistics and transportation is more necessary than ever, and constitutes a critical factor for the evolution of an economically and environmentally stable world. This chapter has reviewed some of the existing literature related to three major and applicable streams of the VRPs, namely, EVRP, GVRP, and HVRP. The main objective of this chapter was to organize the available literature so as to provide a reference point for future research on new aspects of routing problems. A wide range of these three variants and their complicate planning efforts have been studied. We have presented a comprehensive and up-to-date review of the existing studies by paying special attention to environmental issues. The following conclusions and future research perspectives are identified:

## 1.5.1   Future Research Directions in EVRP

Although there is a large literature related to the EVRP scheme, the EV network modeling study is limited and evolving. On this basis, the following research directions are proposed:

   We believe that promising research avenues lie in the integration of additional features in the EVRP. For example, Jie et al. [86] did not consider all realistic situations in the problem. Future work can extend their model by considering the battery recharging time and customer service time window. The capacity limitation of satellites and BSSs can be incorporated also into the manuscript. Another interesting future work could be the development of the rewards-driven mechanism [70] or maintenance operation [46] in the field of EVRP, which can bring greater benefits to logistics companies. One natural extension could be to incorporate in-route recharging into the problem to render it applicable in mid- and long-haul contexts.

   Handling the variation of energy consumption depending on different external factors is one of the most important challenges. Future studies may consider more practical factors related to the ECR function, such as the air conditioner status and environmental temperature. The development of more efficient solution approaches would also be welcome to solve large-sized instances with multiple vehicles. The development of new optimization and hybrid optimization-simulation methods to efficiently cope with the challenges of the incorporation of EVs in the road distribution activities, including dynamic scenarios especially with other types of uncertainty sets would be high value for promoting the desirable shift towards more sustainable energy sources in the logistics and transportation area.

## 1.5.2   Future Research Directions in GVRP

Based on the review on GVRP presented above, we draw the following conclusions about the trends of GVRP, through the analysis of how the GVRP can interact with the traditional VRP variants. There exist extensive VRP variants that can be combined with the GVRP model and make it comprehensive and closer and more applicable to real-world problems. As the fuel consumption model is closely related to the condition of the vehicle, the flexibility offered by using different types of vehicles may result in more reduction of fuel consumption. The behavior of heterogeneous vehicles under varying speed environment is not still explored in the existing literature. Additional constraints like road conditions, vehicle load can be included with varying speed environment in fuel-efficient GVRP and their impact on route cost and fuel consumption can be studied. Also, the researches can be further extended to study the impact on route cost and fuel consumption in routing problems where speed variation can occur with partial and fully charged batteries. The model can also be used to study the impact of speed and fuel consumption with hybrid vehicles, when there is a switch over between battery and fuel which affects the speed.

Due to its significant practical relevance, the modeling of capacity restrictions at the recharging stations is another aspect that has gained increasing interest in the recent literature (see, e.g., [203, 228]). For instance, Upchurch et al. [203] developed the capacitated flow refueling location model that additionally integrates limited capacities for stations to limit the number of vehicles refueled at the same station. The approach of Chung and Kwon [36] considers the fact that the installation of a network of stations is a long-term process that usually requires several periods to be completed. Hence, the model restricts the number of newly established stations per period while pursuing maximization of possible traffic over all periods.

Since long waiting times for recharging would cause considerable customer inconvenience and, as a direct consequence, may even endanger the market adoption of electric cars, its modeling is essential. However, it is worth mentioning that this extension substantially changes the basic problem structure. Note that in the original refueling station location problem model, the coverage of demand depends only on the set of opened stations, but is not influenced by other demands. In the capacitated case, however, demands to be covered may compete for opened stations since their usage on necessary paths is limited. However, the development of algorithms that are based on sophisticated decompositions may, also, in this case, be a promising field of future research.

In the recharging problems, some restrictions in the real world have not yet been accommodated in the routing model. For example, the availability and the fuel capacity of the recharging stations will cause some change to the optimal routes. The stochastic service time of the recharging stations is also worth attention as it influences the time traveled and the arrival time at each point. An immediate extension of the GVRP is to model with techniques like queuing models that seem suitable for tackling the service time problem in this context.

Similar to the travel times, the service times of the customers may also vary depending on the time of the day according to several factors such as ease of access from the road and availability of parking spaces. We believe that the service time, especially in the stochastic form, is not negligible in future research and the real-time transportation information can lay a solid foundation for continued research into the PRP by providing dynamic real-world data. With the support of real-time information about traffic conditions, vehicles can be directed to other roads which are less congested areas. This implies a more environmental-friendly case because less emission is generated when vehicles are traveling at the best speeds to meet service time exactly.

## 1.5.3   Future Research Directions in HVRP

Utilizing HVRP can help companies or organizations reduce their total travel cost by determining the optimal route that serves more consumers in less dependence on fossil fuels. Future developments in this field could address the definition of extensions of this problem in which traffic limited zones, in which it is forbidden to use traditional fuel propulsion are considered. Other developments could address the possibility of partial battery recharging and the usage of the so-called regenerative braking which allows EVs to recover a percentage of their battery charge while traveling on downhill roads. Moreover, multiple recharging technologies, with different recharging times, costs, and availability could be considered, as in the work by Felipe et al. [53] on the pure electric propulsion case.

## References

1. Abdulaal, A., Cintuglu, M.H., Asfour, S., Mohammed, O.A.: Solving the multivariant EV routing problem incorporating V2G and G2V options. IEEE Trans. Transp. Electrification **3**, 238–248 (2017)
2. Adler, J.D., Mirchandani, P.B.: Online routing and battery reservations for electric vehicles with swappable batteries. Transp. Res. Part B Method. **70**, 285–302 (2014)
3. Afroditi, A., Boile, M., Theofanis, S., Sdoukopoulos, E., Margaritis, D.: Electric vehicle routing problem with industry constraints: trends and insights for future research. Transp. Res. Procedia. **3**, 452–459 (2014)
4. Afshar-Bakeshloo, M., Mehrabi, A., Safari, H., Maleki, M., Jolai, F.: A green vehicle routing problem with customer satisfaction criteria. J. Ind. Eng. Int. **12**, 529–544 (2016)
5. Aggoune-Mtalaa, W., Habbas, Z., Ouahmed, A.A., Khadraoui, D.: Solving new urban freight distribution problems involving modular electric vehicles. IET Intell. Transp. Syst. **9**, 654–661 (2015)
6. Amjad, S., Neelakrishnan, S., Rudramoorthy, R.: Review of design considerations and technological challenges for successful development and deployment of plug-in hybrid electric vehicles. Renew. Sustain. Energy Rev. **14**, 1104–1110 (2010)
7. Andelmin, J., Bartolini, E.: An exact algorithm for the green vehicle routing problem. Transp. Sci. **51**, 1288–1303 (2017)
8. Apaydin, O., Gonullu, M.T.: Emission control with route optimization in solid waste collection process: a case study. Sadhana **33**, 71–82 (2008)
9. Arslan, O., Yildiz, B., Karasan, O.E.: Minimum cost path problem for plug-in hybrid electric vehicles. Transp. Res. E Logist. Transp. Rev. **80**, 123–141 (2015)
10. Barco, J., Guerra, A., Muñoz, L., Quijano, N.: Optimal routing and scheduling of charge for electric vehicles: Case study. Working paper. Universidad de los Andes, Bogotá, Colombia (2013)

11. Barco, J., Guerra, A., Muñoz, L., Quijano, N.: Optimal routing and scheduling of charge for electric vehicles: a case study. Math. Probl. Eng. **2017**, 1–16 (2017)
12. Bard, J.F., Huang, L., Jaillet, P., Dror, M.: A decomposition approach to the inventory routing problem with satellite facilities. Transp. Sci. **32**, 189–203 (1998)
13. Barth, M., Younglove, T., Scora, G.: Development of a heavy-duty diesel modal emissions and fuel consumption model. UC Berkeley: California Partners for Advanced Transportation Technology (2005). https://escholarship.org/uc/bibitem/67f0v3zf
14. Basso, R., Kulcsár, B., Egardt, B., Lindroth, P., Sanchez-Diaz, I.: Energy consumption estimation integrated into the Electric Vehicle Routing Problem. Transp. Res. Part D. **69**, 141–167 (2019)
15. Bauer, J., Bektas, T., Crainic, T.G.: Minimizing greenhouse gas emissions in intermodal freight transport: an application to rail service design. J. Oper. Res. Soc. **61**, 530–542 (2010)
16. Beiter, P., Tian, T.: 2015 Renewable Energy Data Book. National Renewable Energy Laboratory. No. DOE/GO-102016-4904 (2016)
17. Bektas, T., Laporte, G.: The pollution-routing problem. Transp. Res. Part B. **45**, 1232–1250 (2011)
18. Bektas, T., Demir, E., Laporte, G.: Green vehicle routing. In: Psaraftis, N.H. (ed.) Green Transportation Logistics: The Quest for Win-Win Solutions, pp. 243–265. Springer International Publishing, Cham (2016)
19. Borcinova, Z.: Two models of the capacitated vehicle routing problem. Croat. Oper. Res. Rev. **8**, 463–469 (2017)
20. Bradley, T.H., Frank, A.A.: Design, demonstrations and sustainability impact assessments for plug-in hybrid electric vehicles. Renew. Sustain. Energy Rev. **13**, 115–128 (2009)
21. Braekers, K., Ramaekersa, K., Nieuwenhu, I.: The vehicle routing problem: state of the art classification and review. Comput. Ind. Eng. **99**, 300–313 (2016)
22. Breunig, U., Baldacci, R., Hartl, R.F., Vidal, T.: The electric two-echelon vehicle routing problem. Technical Report (2018)
23. Breunig, U., Baldacci, R., Hartl, R.F., Vidal, T.: The electric two-echelon vehicle routing problem. Comput. Oper. Res. **103**, 198–210 (2019)
24. Bruglieri, M., Pezzella, F., Pisacane, O., Suraci, S.: A matheuristic for the electric vehicle routing problem with time windows. arXiv:1506.00211 (2015)
25. Bruglieri, M., Pezzella, F., Pisacane, O., Suraci, S.: A variable neighborhood search branching for the electric vehicle routing problem with time windows. Electron. Notes Discrete Math. **47**, 221–228 (2015)
26. Bruglieri, M., Mancini, S., Pezzella, F., Pisacane, O.: A new mathematical programming model for the green vehicle routing problem. Electron. Notes Discrete Math. **55**, 89–92 (2016)
27. Bruglieri, M., Mancini, S., Pezzella, F., Pisacane, O.: The green vehicle routing problem with capacitated alternative fuel stations. Verolog 2017, Amsterdam, 10th–2th July 2017
28. Bruglieri, M., Mancini, S., Pezzella, F., Pisacane, O., Suraci, S.: A three-phase matheuristic for the time-effective electric vehicle routing problem with partial recharges. Electron. Notes Discrete Math. **58**, 95–102 (2017)
29. Bruglieri, M., Mancini, S., Pisacane, O.: Solving the green vehicle routing problem with capacitated alternative fuel stations. In: Proceedings of $16^{th}$ Cologne-Twente Workshop on Graphs and Combinatorial Optimization, Paris, pp. 196–199 (2018)
30. Bruglieri, M., Mancini, S., Pezzella, F., Pisacane, O.: A path-based solution approach for the green vehicle routing problem. Comput. Oper. Res. **103**, 109–122 (2019)
31. Casals, L.C., Martinez-Laserna, E., García, B.A., Nieto, N.: Sustainability analysis of the electric vehicle use in Europe for CO2 emissions reduction. J. Clean. Prod. **127**, 425–437 (2016)
32. Çatay, B., Keskin, M.: The impact of quick charging stations on the route planning of electric vehicles. 2017 IEEE Symposium on Computers and Communications (ISCC), Heraklion (2017)
33. Chan, C.: The state of the art of electric, hybrid, and fuel cell vehicles. IEEE **95**, 704–718 (2007)

34. Chen, J., Qi, M., Miao, L.: The electric vehicle routing problem with time windows and battery swapping stations. Paper Presented at 2016 IEEE International Conference on the Industrial Engineering and Engineering Management (IEEM), pp. 712–716 (2016)
35. Christofides, N.: The vehicle routing problem. Revue française d'automatique, informatique, recherche opérationnelle. Rech. opér. **10**, 55–70 (1976)
36. Chung, S.H., Kwon, C.: Multi-period planning for electric car charging station locations: a case of Korean expressways. Eur. J. Oper. Res. **242**, 677–687 (2015)
37. Çimen, M., Soysal, M.: Time-dependent green vehicle routing problem with stochastic vehicle speeds: an approximate dynamic programming algorithm. Transp. Res. D Transp. Environ. **54**, 82–98 (2017)
38. Cirovic, G., Pamuočar, D., Božanic, D.: Green logistic vehicle routing problem: routing light delivery vehicles in urban areas using a neuro-fuzzy model. Expert Syst. Appl. **41**, 4245–4258 (2014)
39. Conrad, R.G., Figliozzi, M.A.: The recharging vehicle routing problem. In: Doolen, T., Van Aken, E. (eds.) Proceedings of the 2011 Industrial Engineering Research Conference, Reno, Nevada (2011)
40. Dabia, S., Demir, E., Van Woensel, T.: An exact approach for the pollution-routing problem. Transp. Sci. **51**, 607–628 (2016)
41. Demir, E., Bektas, T., Laporte, G.: An adaptive large neighborhood search heuristic for the pollution-routing problem. Eur. J. Oper. Res. **223**, 346–359 (2012)
42. Demir, E., Bektaş, T., Laporte, G.: The bi-objective pollution-routing problem. Eur. J. Oper. Res. **232**, 464–478 (2014)
43. Demir, E., Bektaş, T., Laporte, G.: A review of recent research on green road freight transportation. Eur. J. Oper. Res. **237**, 775–793 (2014)
44. Desaulniers, G., Errico, F., Irnich, S., Schneider, M.: Exact algorithms for electric vehicle-routing problems with time windows. Oper. Res. **64**, 1388–1405 (2016)
45. Ding, N., Batta, R., Kwon, C.: Conflict-Free Electric Vehicle Routing Problem with Capacitated Charging Stations and Partial Recharge. SUNY, Buffalo (2015)
46. Duan, C.Q., Deng, C., Gharaei, A., Wu, J., Wang, B.R.: Selective maintenance scheduling under stochastic maintenance quality with multiple maintenance actions. Int. J. Prod. Res. **56**, 7160–7178 (2018)
47. Dukkanci, O., Kara, B.Y., Bektaş T.: The green location-routing problem. Comput. Oper. Res. **105**, 187–202 (2019)
48. Erdogan, S., Miller-Hooks, E.: A green vehicle routing problem. Transp. Res. E Logist. Transp. Rev. **48**, 100–114 (2012)
49. Eshtehadi, R., Fathian, M., Demir, E.: Robust solutions to the pollution-routing problem with demand and travel time uncertainty. Transp. Res. Transp. Environ. **51**, 351–363 (2017)
50. Fagerholt, K.: Optimal fleet design in a ship routing problem. Int. Trans. Oper. Res. **6**, 453–464 (1999)
51. Fagerholt, K., Laporte, G., Norstad, I.: Reducing fuel emissions by optimizing speed on shipping routes. J. Oper. Res. Soc. **61**, 523–529 (2010)
52. Faulin, J., Juan, A., Lera, F., Grasman, S.: Solving the capacitated vehicle routing problem with environmental criteria based on real estimations in road transportation: a case study. Procedia—Soc. Behav. Sci. **20**, 323–334 (2011)
53. Felipe, A., Ortuño, M.T., Righini, G., Tirado, G.: A heuristic approach for the green vehicle routing problem with multiple technologies and partial recharges. Transp. Res. E Logist. Transp. Rev. **71**, 111–128 (2014)
54. Feng, W., Figliozzi, M.: An economic and technological analysis of the key factors affecting the competitiveness of electric commercial vehicles: a case study from the USA market. Transp. Res. C Emerg. Technol. **26**, 135–145 (2013)
55. Fernandez, R. A.: A more realistic approach to electric vehicle contribution to greenhouse gas emissions in the city. J. Clean. Prod. **172**, 949–959 (2018)
56. Ferro, G., Paolucci, M., Robba, M.: An optimization model for electrical vehicles routing with time of use energy pricing and partial recharging. IFACPapersOnLine **51**, 212–217 (2018)

57. Fetene, M.G., Kaplan, S., Mabit, L.S., Jensen, F.A., Prato, G.C.: Harnessing big data for estimating the energy consumption and driving range of electric vehicles. Transp. Res. Transp. Environ. **54**, 1–11 (2017)
58. Figliozzi, M.A.: An iterative route construction and improvement algorithm for the vehicle routing problem with soft time windows. Transp. Res. Part C Emerg. Technol. **18**, 668–679 (2010)
59. Fiori, C., Ahn, K., Rakha, A.H.: Power-based electric vehicle energy consumption model: model development and validation. Appl. Energy **168**, 257–268 (2016)
60. Fontana, M.W.: Optimal routes for electric vehicles facing uncertainty, congestion, and energy constraints. Ph.D. thesis. Massachusetts Institute of Technology, Massachusetts (2013)
61. Franceschetti, A., Honhon, D., Van Woensel, T., Bektas, T., Laporte, G.: The time dependent pollution-routing problem. Transp. Res. B Methodol. **56**, 265–293 (2013)
62. Franceschetti, A., Demir, E., Honhon, D., Van Woensel, T., Laporte, G., Stobbe, M.: A metaheuristic for the time-dependent pollution-routing problem. Eur. J. Oper. Res. **259**, 972–991 (2017)
63. Frank, S., Preis, H., Nachtigall, K.: On the modeling of recharging stops in context of vehicle routing problems. In: Operations Research Proceedings 2013, pp. 129–135. Springer, Berlin (2014)
64. Froger, A., Mendoza, J.E., Jabali, O., Laporte, G.: A matheuristic for the electric vehicle routing problem with capacitated charging stations. Technical Report (2017)
65. Froger, A., Mendoza, J., Jabali, O., Laporte, G.: New formulations for the electric vehicle routing problem with nonlinear charging functions. Technical Report, CIRRELT (2017)
66. Froger, A., Mendoza, J.E., Jabali, O., Laporte, G.: Improved formulations and algorithmic components for the electric vehicle routing problem with nonlinear charging functions. Comput. Oper. Res. **104**, 256–294 (2019)
67. Fukasawa, R., Longo, H., Lysgaard, J., de Aragão, M.P., Reis, M., Uchoa, E., Werneck, R.F.: Robust branch-and-cut-and-price for the capacitated vehicle routing problem. Math. Program. **106**, 491–511 (2006)
68. Fukasawa, R., He, Q., Song, Y.: A branch-cut-and-price algorithm for the energy minimization vehicle routing problem. Transp. Sci. **50**, 23–34 (2015)
69. Fukasawa, R., He, Q., Song, Y.: A disjunctive convex programming approach to the pollution-routing problem. Transp. Res. Part B. **94**, 61–79 (2016)
70. Gharaei, A., Naderi, B., Mohammadi, M.: Optimization of rewards in single machine scheduling in the rewards-driven systems. Manag. Sci. Lett. **5**, 629–638 (2015)
71. Goeke, D., Schneider, M.: Routing a mixed fleet of electric and conventional vehicles. Eur. J. Oper. Res. **245**, 81–99 (2015)
72. Golden, B.L., Raghavan, S., Wasil, E.: The Vehicle Routing Problem: Latest Advances and New Challenges. Springer, New York (2008)
73. Gonçalves, F., Cardoso, S.R., Relvas, S., Barbosa-Povoa, A.: Optimization of a distribution network using electric vehicles: a VRP problem. In: Proceedings of the IO2011-15 Congresso da associação Portuguesa de Investigação Operacional, Coimbra, pp. 18–20 (2011)
74. Gong, Q., Li, Y., Peng, Z.R.: Trip-based optimal power management of plug-in hybrid electric vehicles. IEEE Trans. Veh. Technol. **57**, 3393–3401 (2008)
75. Göpfert, P., Bock, S.: A Branch&Cut approach to recharging and refueling infrastructure planning. Eur. J. Oper. Res. **279**, 808–823 (2019)
76. Grandinetti, L., Guerriero, F., Pezzella, F., Pisacane, O.: A pick-up and delivery problem with time windows by electric vehicles. Int. J. Prod. Qual. Manag. **18**, 403–423 (2016)
77. Hannan, M.A., Azidin, F.A., Mohamed, A.: Hybrid electric vehicles and their challenges: a review. Renew. Sustain. Energy Rev. **29**, 135–150 (2014)
78. Hiermann, G., Puchinger, J., Hartl, R.F.: The electric fleet size and mix vehicle routing problem with time windows and recharging stations. Eur. J. Oper. Res. **252**, 995–1018 (2016)
79. Hiermann, G., Puchinger, J., Ropke, S., Hartl, R.F.: The electric fleet size and mix vehicle routing problem with time windows and recharging stations. Eur. J. Oper. Res. **252**, 995–1018 (2016)

80. Hiermann, G., Hartl, R.F., Puchinger, J., Vidal, T.: Routing a mix of conventional, plug-in hybrid, and electric vehicles. Eur. J. Oper. Res. **272**, 235–248 (2019)
81. Hof, J., Schneider, M., Goeke, D.: Solving the battery swap station location-routing problem with capacitated electric vehicles using an AVNS algorithm for vehicle-routing problems with intermediate stops. Transp. Res. B Methodol. **97**, 102–112 (2017)
82. Iwata, K., Matsumoto, S.: Use of hybrid vehicles in Japan: an analysis of used car market data. Transp. Res. Part D. **46**, 200–206 (2016)
83. Jabali, O., Van Woensel, T., de Kok, A.: Analysis of travel times and $CO_2$ emissions in time-dependent vehicle routing. Prod. Oper. Manag. **21**, 1060–1074 (2012)
84. Jabir, E., Panicker, V., Sridharan, R.: Design and development of a hybrid ant colony-variable neighbourhood search algorithm for a multi-depot green vehicle routing problem. Transp. Res. Transp. Environ. **57**, 422–457 (2017)
85. Jemai, J., Zekri, M., Mellouli, K.: An NSGA-II algorithm for the green vehicle routing problem. In: Hao, J.-K., Middendorf, M. (eds.) Proceedings of the 12th European Conference on Evolutionary Computation in Combinatorial Optimization, EvoCOP 2012. Lecture Notes in Computer Science, vol. 7245, pp. 37–48. Springer, Málaga (2012)
86. Jie, W., Yang, J., Zhang, M., Huang, Y.: The two-echelon capacitated electric vehicle routing problem with battery swapping stations: formulation and efficient methodology. Eur. J. Oper. Res. **272**, 879–904 (2019)
87. Jovanovic, A.D., Pamucar, D.S., Pejcic-Tarle, S.: Green vehicle routing in urban zones—a neuro-fuzzy approach. Expert Syst. Appl. **41**, 3189–3203 (2014)
88. Jovicic, N.M., Boškovic, G.B., Vujic, G.V., Jovicic, G.R., Despotovic, M.Z., Milovanovic, D.M., et al.: Route optimization to increase energy efficiency and reduce fuel consumption of communal vehicles. Therm. Sci. **14**, 67–78 (2010)
89. Juan, A., Goentzel, J., Bektas, T.: Routing fleets with multiple driving ranges: is it possible to use greener fleet configurations? Appl. Soft Comput. **21**, 84–94 (2014)
90. Keskin, M., Çatay, B.: Partial recharge strategies for the electric vehicle routing problem with time windows. Transp. Res. C Emerg. Technol. **65**, 111–127 (2016)
91. Keskin, M., Çatay, B.: A matheuristic method for the electric vehicle routing problem with time windows and fast chargers. Comput. Oper. Res. **100**, 172–188 (2018)
92. Keskin, M., Laporte, G., Çatay, B.: Electric vehicle routing problem with time dependent waiting times at recharging stations. In: Odysseus 2018: 7th International Workshop on Freight Transportation and Logistics, pp. 1–4 (2018)
93. Keskin, M., Laporte G., Çatay, B.: Electric vehicle routing problem with time-dependent waiting times at recharging stations. Comput. Oper. Res. **107**, 77–94 (2019)
94. Koç, C., Karaoglan, I.: The green vehicle routing problem: a heuristic based exact solution approach. Appl. Soft Comput. **39**, 154–164 (2016)
95. Koç, Ç., Bektas, T., Jabali, O., Laporte, G.: The fleet size and mix pollution-routing problem. Transp. Res. Part B. **70**, 239–254 (2014)
96. Koç, C., Jabali, O., Laporte, G.: Long-haul vehicle routing and scheduling with idling options. J. Oper. Res. Soc. **69**, 235–246 (2018)
97. Kopfer, H., Vornhusen, B.: Energy vehicle routing problem for differently sized and powered vehicles. J. Bus. Econ. 1–29 (2018)
98. Kopfer, H., Schnberger, J., Kopfer, H.: Reducing greenhouse gas emissions of a heterogeneous vehicle fleet. Flex. Serv. Manuf. J. **26**, 221–248 (2014)
99. Kullman, N.D., Goodson, J.C., Mendoza, J.E.: Dynamic electric vehicle routing: heuristics and dual bounds (2018). Working paper or preprint. https://hal.archives-ouvertes.fr/hal-01928730
100. Lebeau, P., De Cauwer, C., Van Mierlo, J., Macharis, C., Verbeke, W., Coosemans, T.: Conventional, hybrid, or electric vehicles: which technology for an urban distribution centre. Sci. World J. 302867 (2015)

101. Leggieri, V., Haouari, M.: A practical solution approach for the green vehicle routing problem. Transp. Res. E Logist. Transp. Rev. **104**, 97–112 (2017)
102. Li, J.: Vehicle routing problem with time windows for reducing fuel consumption. J. Comput. **7**, 3020–3027 (2012)
103. Li, Y., Zhan, C., Jong, de M., Lukszo, Z.: Business innovation and government regulation for the promotion of electric vehicle use: lessons from Shenzhen, China. J. Clean. Prod. **134**, 371–383 (2016)
104. Li, C., Ding, T., Liu, X., Huang, C.: An electric vehicle routing optimization model with hybrid plug-in and wireless charging systems. IEEE **6**, 27569–27578 (2018)
105. Li, J., Wang, D., Zhang, J.: Heterogeneous fixed fleet vehicle routing problem based on fuel and carbon emissions. J. Clean. Prod. **201**, 896–908 (2018)
106. Li, Y., Soleimani, H., Zohal, M.: An improved ant colony optimization algorithm for the multi-depot green vehicle routing problem with multiple objectives. J. Clean. Prod. **227**, 1161–1172 (2019)
107. Li, X., Zhao, Y., Shi, X., Shi, T.: A hybridization of memetic algorithm with SVND for solving a hybrid vehicle routing problem. In: Proceedings of the 2019 4th International Conference on Mathematics and Artificial Intelligence, Chegndu, pp. 162–166 (2019)
108. Liao, C.S., Lu, S.H., Shen, Z.J.M.: The electric vehicle touring problem. Transp. Res. Part B Methodol. **86**, 163–180 (2016)
109. Liu, L., Liu, C., Liu, X., Wang, S., Zhou, W., Zhan, Z.: Research and application of multiple constrained hot strip mill scheduling problem based on HPSA. Int. J. Adv. Manuf. Technol. **81**, 1817–1829 (2015)
110. Liu, K., Yamamoto, T., Morikawa, T.: Impact of road gradient on energy consumption of electric vehicles. Transp. Res. Part D **54**, 74–81 (2017)
111. Li-Ying, W., Yuan-Bin, S.: Multiple charging station location-routing problem with time window of electric vehicle. J. Eng. Sci. Technol. Rev. **8**, 190–201 (2015)
112. Lin, C., Choy, K.L., Ho, G.T.S., Chung, S.H., Lam, H.Y.: Survey of green vehicle routing problem: past and future trends. Expert Syst. Appl. **41**, 1118–1138 (2014)
113. Lin, J., Zhou, W., Wolfson, O.: Electric vehicle routing problem. Transp. Res. Procedia. **12**, 508–521 (2016)
114. Lu, X., Zhou, K., Yang, S., Liu, H.: Multi-objective optimal load dispatch of microgrid with stochastic access of electric vehicles. J. Clean. Prod. **195**, 187–199 (2018)
115. Lübbecke, M.E., Desrosiers, J.: Selected topics in column generation. Oper. Res. **53**, 1007–1023 (2005)
116. Kancharla, S., Ramadurai, G.: Incorporating driving cycle based fuel consumption estimation in green vehicle routing problems. Sustain. Cities Soc. 214–221 (2018)
117. Kara, İ., Kara, B.Y., Yetis, M.K.: Energy minimizing vehicle routing problem. Combinatorial Optimization and Applications: First International Conference, COCOA 4616, 62–71 (2007)
118. Karak A., Abdelghany, K.: The hybrid vehicle-drone routing problem for pick-up and delivery services. Transp. Res. Part C **102**, 427–449 (2019)
119. Kargari Esfand Abad, H., Vahdani, B., Sharifi, M., Etebari, F.: A bi-objective model for pickup and delivery pollution-routing problem with integration and consolidation shipments in crossdocking system. J. Clean. Prod. **193**, 784–801 (2018)
120. Kazemian, I., Aref, S.: A green perspective on capacitated time-dependent vehicle routing problem with time windows. Int. J. Supply Chain Inventory Manag. **2**, 20–38 (2017)
121. Koyuncu, I., Yavuz, M.: Duplicating nodes or arcs in green vehicle routing: a computational comparison of two formulations. Transp. Res. Part E **122**, 605–623 (2019)
122. Kramer, R., Maculan, N., Subramanian, A., Vidal, T.: A speed and departure time optimization algorithm for the pollution-routing problem. Eur. J. Oper. Res. **247**, 782–787 (2015)
123. Kramer, R., Subramanian, A., Vidal, T., Cabral, L.D.A.F.: A matheuristic approach for the pollution-routing problem. Eur. J. Oper. Res. **243**, 523–539 (2015)
124. Küçükoglu, I., Öztürk, N.: Route optimization of the electric vehicles with heterogeneous fleet. CBU J. Sci. **12**, 525–533 (2016)

125. Küçükoglu, I., Dewil, R., Cattrysse, D.: Hybrid simulated annealing and tabu search method for the electric travelling salesman problem with time windows and mixed charging rates. Expert Syst. Appl. **134**, 279–303 (2019)

126. Kuo, Y.: Using simulated annealing to minimize fuel consumption for the time-dependent vehicle routing problem. Comput. Ind. Eng. **59**, 157–165 (2010)

127. Kwon, Y.J., Choi, Y.J., Lee, D.H.: Heterogeneous fixed fleet vehicle routing considering carbon emission. Transp. Res. Part D: Transp. Environ. **23**, 81–89 (2013)

128. Macrina, G., Di Puglia Pugliese, L., Guerriero, F., Laporte, G.: The green mixed fleet vehicle routing problem with partial battery recharging and time windows. Comput. Oper. Res. **101**, 183–199 (2019)

129. Macrina, G., Laporte, G., Guerriero, F., Luigi, P., Pugliese, D.: An energy-efficient green-vehicle routing problem with mixed vehicle fleet, partial battery recharging and time windows. Eur. J. Oper. Res. **276**, 971–982 (2019)

130. Madankumar, S., Rajendran, C.: Mathematical models for green vehicle routing problems with pickup and delivery: a case of semiconductor supply chain. Comput. Oper. Res. **89**, 183–192 (2018)

131. Maden, W., Eglese, R., Black, D.: Vehicle routing and scheduling with timevarying data: a case study. J. Oper. Res. Soc. **61**, 515–522 (2010)

132. Mahmudul, H.M., Hagos, F.Y., Mamat, R., Adam, A.A., Ishak, W.F.W., Alenezi, R.L.: Production, characterization and performance of biodiesel as an alternative fuel in diesel engines–a review. Renew. Sustain. Energy Rev. **72**, 497–509 (2017)

133. Majidi, S., Hosseini-Motlagh, S.M., Ignatius, J.: Adaptive large neighborhood search heuristic for pollution-routing problem with simultaneous pickup and delivery. Soft Comput. **22**, 2851–2865 (2018)

134. Mancini, S.: The hybrid vehicle routing problem. Transp. Res. Part C Emerg. Technol. **78**, 1–12 (2017)

135. Maraš, V.: Determining optimal transport routes of inland waterway container ships. Transp. Res. Rec. **2026**, 50–58 (2008)

136. Masmoudi, M.A., Hosny, M., Demir, E., Genikomsakis, K.N., Cheikhrouhou, N.: The dial-a-ride problem with electric vehicles and battery swapping stations. Transp. Res. E Logist. Transp. Rev. **118**, 392–420 (2018)

137. McKinnon, A.: $CO_2$ Emissions from freight transport in the UK. Technical Report. Prepared for the Climate Change Working Group of the Commission for Integrated Transport, London (2007)

138. Micheli, G.J.L., Mantella, F.: Modelling an environmentally-extended inventory routing problem with demand uncertainty and a heterogeneous fleet under carbon control policies. Int. J. Prod. Econ. **204**, 316–327 (2018)

139. Mirmohammadi, S.H., Babaee Tirkolaee, E., Goli, A., Dehnavi-Arani, S.: The periodic green vehicle routing problem with considering of time-dependent urban traffic and time windows. Iran Univ. Sci. Technol. **7**, 143–156 (2017)

140. Montoya, A., Guéret, C., Mendoza, J.E., Villegas, J.G.: A multi-space sampling heuristic for the green vehicle routing problem. Transp. Res. C Emerg. Technol. **70**, 113–128 (2016)

141. Montoya, A., Guéret, C., Mendoza, J.E., Villegas, J.G.: The electric vehicle routing problem with nonlinear charging function. Transp. Res. B Methodol. **103**, 87–110 (2017)

142. Montoya-Torres, J.R., López Franco, J., Nieto Isaza, S., Felizzola Jiménez, H., Herazo-Padilla, N.: A literature review on the vehicle routing problem with multiple depots. Comput. Ind. Eng. **79**, 115–129 (2015)

143. Murgovski, N., Johannesson, L., Sjöberg, J.: Engine on/off control for dimensioning hybrid electric powertrains via convex optimization. IEEE Trans. Vehicular Tech. **62**, 2949–2962 (2013)

144. Nakata, T.: Analysis of the impact of hybrid vehicles on energy systems in Japan. Transp. Res. Part D. **5**, 373–383 (2000)

145. Nanthavanij, S., Boonprasurt, P., Jaruphongsa, W., Ammarapala, V.: Vehicle routing problem with manual materials handling: flexible delivery crewvehicle assignments. In: Proceeding of the 9th Asia Pacific Industrial Engineering and Management System Conference, Nusa Dua, Bali (2008)
146. Nejad, M., Mashayekhy, L., Grosu, D., Chinnam, R.: Optimal routing for plug-in hybrid electric vehicles. Transp. Sci. **51**, 1304–1325 (2016)
147. Niu, Y., Yang, Z., Chen, P., Xiao, J.: Optimizing the green open vehicle routing problem with time windows by minimizing comprehensive routing cost. J. Clean. Prod. **171**, 962–971 (2018)
148. Ouahmed, A. A., Aggoune-Mtalaa, W., Habbas, Z., Khadraoui, D.: eM-VRP: A new class of vehicle routing problem based on a new concept of modular electric vehicle. Paper Presented at the Transport Research Arena (TRA) Transport Solutions: From Research to Deployment (2014)
149. Palmer, A.: The development of an integrated routing and carbon dioxide emissions model for goods vehicles. Ph.D. Dissertation, School of Management, Cranfield University (2007)
150. Pan, S., Ballot, E., Fontane, F.: The reduction of greenhouse gas emissions from freight transport by pooling supply chains. Int. J. Prod. Econ. **143**, 86–94 (2013)
151. Park, Y., Chae, J.: A review of the solution approaches used in recent G-VRP (Green Vehicle Routing Problem). Int. J. Adv. Logist. **3**, 27–37 (2014)
152. Paz, J., Granada-Echeverri, M., Escobar, J.: The multi-depot electric vehicle location routing problem with time windows. Int. J. Ind. Eng. Comput. **9**, 123–136 (2018)
153. Pelletier, S., Jabali, O., Laporte, G.: Goods distribution with electric vehicles: review and research perspectives. Technical Report CIRRELT-2014-44, CIRRELT, Montréal (2014)
154. Pelletier, S., Jabali, O., Laporte, G.: Goods distribution with electric vehicles: review and research perspectives. Transp. Sci. **50**, 3–22 (2016)
155. Pelletier, S., Jabali, O., Laporte, G.: Charge scheduling for electric freight vehicles. CIRRELT Technical Report, vol. 20, pp. 17–37 (2017)
156. Pelletier, S., Jabali, O., Laporte, G., Veneroni, M.: Battery degradation and behavior for electric vehicles: review and numerical analyses of several models. Transp. Res. B Methodol. **103**, 158–187 (2017)
157. Pelletier, S., Jabali, O., Laporte, G.: The electric vehicle routing problem with energy consumption uncertainty. Transp. Res. Part B. **126**, 225–255 (2019)
158. Penna, P.H.V., Afsar, H.M., Prins, C., Prodhon, C.: A hybrid iterative local search algorithm for the electric fleet size and mix vehicle for the electric fleet size and mix vehicle routing problem with time windows and recharging stations. IFAC-PapersOnLine **49**, 955–960 (2016)
159. Pisacane, O., Bruglieri, M., Mancini, S., Pezzella, F.: A path-based mixed integer linear programming formulation for the green vehicle routing problem. In: Proceedings of the Annual Workshop of the EURO Working Group on Vehicle Routing and Logistics (VeRoLog), Amsterdam, 10–12 July 2017
160. Pitera, K., Sandoval, F., Goodchild, A.: Evaluation of emissions reduction in urban pickup systems. Transp. Res. Rec. J. Transp. Res. Board **2224**, 8–16 (2011)
161. Plotkin, S., Santini, D., Vyas, A., Anderson, J., Wang, M., Bharathan, D., He, J.: Hybrid electric vehicle technology assessment: Methodology, analytical issues, and interim results. Technical report, Argonne National Lab, Lemont (2002). https://doi.org/10.2172/807353
162. Poonthalir, G., Nadarajan, R.: A fuel efficient green vehicle routing problem with varying speed constraint (F-GVRP). Expert Syst. Appl. **100**, 131–144 (2018)
163. Poonthalir, G., Nadarajan, R.: Green vehicle routing problem with queues. In: Proceeding of Expert Systems with Applications (2019)
164. Preis, H., Frank, S., Nachtigall, K.: Energy-optimized routing of electric vehicles in urban delivery systems. In: Helber, S., Breitner, M., Rösch, D., Schön, C., von der Schulenburg, J.M.G., Sibbertsen, P., Steinbach, M., Weber, S., Wolter, A. (eds.) Proceedings of the Operations Research, pp. 583–588. Springer International Publishing, Cham (2014)
165. Pronello, C., André, M.: Pollutant emissions estimation in road transport models. INRETS-LTE Report 2007 (2000)

166. Qazvini, Z., Ebrahimi, A., Mohsen, S., Mina, H.: A green multi-depot location routing model with split-delivery and time window. Int. J. Manag. Conc. Philos. **9**, 271–282 (2016)
167. Qian, J., Eglese, R.: Fuel emissions optimization in vehicle routing problems with time-varying speeds. Eur. J. Oper. Res. **248**, 840–848 (2016)
168. Qiu, Y., Qiao, J., Pardalos, P.M.: A branch-and-price algorithm for production routing problems with carbon cap-and-trade. Omega. **68**, 49–61 (2017)
169. Rabbani, M., Bosjin, S., Yazdanparast, R., Saravi, N.: A stochastic time-dependent green capacitated vehicle routing and scheduling problem with time window, resiliency and reliability: a case study. Decis. Sci. Lett. **7**, 381–394 (2018)
170. Raeesi R., Zografos, K.G.: The multi-objective Steiner pollution-routing problem on congested urban road networks. Transp. Res. Part B. **122**, 457–485 (2019)
171. Rahman, I., Vasant, P.M., Singh, B.S.M., Abdullah-Al-Wadud, M., Adnan, N.: Review of recent trends in optimization techniques for plug-in hybrid, and electric vehicle charging infrastructures. Renew. Sustain. Energy Rev. **58**, 1039–1047 (2016)
172. Rauniyar, A., Nath, R., Muhuri, P.K.: Multi-factorial evolutionary algorithm based novel solution approach for multi-objective pollution-routing problem. Comput. Ind. Eng. **130**, 757–771 (2019)
173. Rezgui, D., Chaouachi-Siala, J., Aggoune-Mtalaa, W., Bouziri, H.: Application of a memetic algorithm to the fleet size and mix vehicle routing problem with electric modular vehicles. Paper Presented at the Proceedings of the Genetic and Evolutionary Computation Conference Companion (2017)
174. Roberti, R., Wen, M.: The electric traveling salesman problem with time windows. Transp. Res. E Logist. Transp. Rev. **89**, 32–52 (2016)
175. Salimifard, K., Shahbandarzaden, H., Raeesi, R.: Green transportation and the role of operation research. In: 2012 International Conference on Traffic and Transportation Engineering, Singapore (2012)
176. Sambracos, E., Paravantis, J.A., Tarantilis, C.D., Kiranoudis, C.T.: Dispatching of small containers via coastal freight liners: the case of the Aegean Sea. Eur. J. Oper. Res. **152**, 365–381 (2004)
177. Sassi, O., Cherif, W.R., Oulamara, A.: Vehicle routing problem with mixed fleet of conventional and heterogenous electric vehicles and time dependent charging costs. Technical Report (2014)
178. Sawik, B, Faulin, J, Elena, P.B.: A multicriteria analysis for the green VRP: a case discussion for the distribution problem of a Spanish retailer. Transp. Res. Proc. **22**, 305–313 (2017)
179. Sbihi, A., Eglese, R.W.: Combinatorial optimization and green logistics. 4OR Q J. Oper. Res. **5**, 99–116 (2007)
180. Schiffer, M., Walther, G.: An adaptive large neighborhood search for the location routing problem with intra-route facilities. Transp. Sci. **52**, 331–352 (2017)
181. Schiffer, M., Walther, G.: The electric location routing problem with time windows and partial recharging. Eur. J. Oper. Res. **260**, 995–1013 (2017)
182. Schiffer, M., Walther, G.: Strategic planning of electric logistics fleet networks: a robust location-routing approach. Omega **80**, 31–42 (2018)
183. Schiffer, M., Schneider, M., Laporte, G.: Designing sustainable mid-haul logistics networks with intra-route multi-resource facilities. Eur. J. Oper. Res. **265**, 517–532 (2018)
184. Schneider, M., Stenger, A., Goeke, G.: The electric vehicle-routing problem with time windows and recharging stations. Transp. Sci. **48**, 500–520 (2014)
185. Schneider, M., Stenger, A., Hof, J.: An adaptive VNS algorithm for vehicle routing problems with intermediate stops. OR Spectr. **37**, 353–387 (2015)
186. Sciarretta A, Guzzella, L.: Control of hybrid electric vehicles. IEEE Control Syst. **27**, 60–70 (2007)

187. Scott, C., Urquhart, N., Hart, E.: Influence of topology and payload on $CO_2$ optimised vehicle routing. In: Chio, C., Brabazon, A., Caro, G., Ebner, M., Farooq, M., Fink, A., Grahl, J., Greenfield, G., Machado, P., O'Neill, M., Tarantino, E., Urquhart, N. (eds.) Applications of Evolutionary Computation. Lecture Notes in Computer Science, vol. 6025, pp. 141–150. Springer, Berlin (2010)

188. Shao, S., Guan, W., Ran, B., He, Z., Bi, J.: Electric vehicle routing problemwith charging time and variable travel time. Math. Probl. Eng. **2017**, 1–13 (2017)

189. Shao, S., Guan, W., Bi, J.: Electric vehicle routing problem with charging demands and energy consumption. IET Intell. Transp. Syst. **12**, 202–212 (2018)

190. Silva, C., Ross, M., Farias, T.: Evaluation of energy consumption, emissions and cost of plug-in hybrid vehicles. Energy Convers. Manag. **50**, 1635–1643 (2009)

191. Soysal, M., Çimen, M.: A simulation based restricted dynamic programming approach for the green time dependent vehicle routing problem. Comput. Oper. Res. **88**, 297–305 (2017)

192. Soysal, M., Bloemhof-Ruwaard, J.M., Bektas, T.: The time-dependent two-echelon capacitated vehicle routing problem with environmental considerations. Int. J. Prod. Econ. **164**, 366–378 (2015)

193. Sun, Z., Zhou, X.: To save money or to save time: intelligent routing design for plug-in hybrid electric vehicle. Transp. Res. Part D. **43**, 238–250 (2016)

194. Suzuki, Y.: A new truck-routing approach for reducing fuel consumption and pollutants emission. Transp. Res. Part D Transp. Environ. **16**, 73–77 (2011)

195. Suzuki, Y.: A dual-objective metaheuristic approach to solve practical pollution routing problem. Int. J. Prod. Econ. **176**, 143–153 (2016)

196. Suzuki, Y., Lan, B.: Cutting fuel consumption of truckload carriers by using new enhanced refueling policies. Int. J. Prod. Econ. **202**, 69–80 (2018)

197. Sweda, T.M., Dolinskaya, I.S., Klabjan, D.: Adaptive routing and recharging policies for electric vehicles. Transp. Sci. **51**, 1326–1348 (2017)

198. Taveares, G., Zaigraiova, Z., Semiao, V., da Graca Carvalho, M.: A case study of fuel savings through optimization of MSW transportation routes. Manag. Environ. Qual. Int. J. **19**, 444–454 (2008)

199. Teng, L., Zhang, Z.: Green vehicle routing problem with load factor. Adv. Transp. Stud. **3**, 75–82 (2016)

200. Tiwari, A., Chang, P.C.: A block recombination approach to solve green vehicle routing problem. Int. J. Prod. Econ. **164**, 379–387 (2015)

201. Toro, E.M., Franco, J.F., Echeverri, M.G., Guimaraes, F.G.: A multi-objective model for the green capacitated location-routing problem considering environmental impact. Comput. Ind. Eng. **110**, 114–125 (2017)

202. Ubeda, S., Arcelus, F.J., Faulin, J.: Green logistics at Eroski: a case study. Int. J. Prod. Econ. **131**, 44–51 (2011)

203. Upchurch, C., Kuby, M., Lim, S.: A model for location of capacitated alternativefuel stations. Geogr. Anal. **41**, 85–106 (2009)

204. Van den Hove, A., Verwaeren, J., Van den Bossche, J., Theunis, J., De Baets, B.: Development of a land use regression model for black carbon using mobile monitoring data and its application to pollution-avoiding routing. Environ. Res. (2019). https://doi.org/10.1016/j.envres.2019.108619

205. Van Duin, J.H.R., Tavasszy, L.A., Quak, H.J.: Towards E(lectric)-urban freight: first promising steps in the electric vehicle revolution. Eur. Transp. Trasp. Eur. **54**, 1–19 (2013)

206. Verma, P., Sharma, M.P.: Review of process parameters for biodiesel production from different feedstocks. Renew. Sustain. Energy Rev. **62**, 1063–1071 (2016)

207. Villegas, J., Guéret, C., Mendoza, J.E., Montoya, A.: The technician routing and scheduling problem with conventional and electric vehicle (2018). https://hal.archives-ouvertes.fr/hal-01813887

208. Wang, H., Cheu, R.: Operations of a taxi fleet for advance reservations using electric vehicles and charging stations. Transp. Res. Rec. J. Transp. Res. Board. **2352**(1), 1–10 (2013)

209. Wang, Y.W., Lin, C.C., Lee, T.J.: Electric vehicle tour planning. Transp. Res. D Transp. Environ. **63**, 121–136 (2018)
210. Wang, J., Yao, S., Sheng, J., Yang, H.: Minimizing total carbon emissions in an integrated machine scheduling and vehicle routing problem. J. Clean. Prod. **229**, 1004–1017 (2019)
211. Wang, Y., Assogba, K., Fan, J., Xu, M., Liu, Y., Wang, H.: Multi-depot green vehicle routing problem with shared transportation resource: integration of time-dependent speed and piecewise penalty cost. J. Clean. Prod. **232**, 12–29 (2019)
212. Wen, M., Linde, E., Ropke, S., Mirchandani, P., Larsen, A.: An adaptive large neighborhood search heuristic for the electric vehicle scheduling problem. Comput. Oper. Res. **76**, 73–83 (2016)
213. Worley, O., Klabjan, D., Sweda, T.M.: Simultaneous vehicle routing and charging station siting for commercial electric vehicles. Paper Presented at the IEEE International Electric Vehicle Conference (IEVC) (2012)
214. Xiao, Y., Konak, A.: Green vehicle routing problem with time-varying traffic congestion. In: Proceedings of the Fourteenth INFORMS Computing Society Conference, pp. 134–148 (2015)
215. Xiao, Y., Zhao, Q., Kaku, I., Xu, Y.: Development of a fuel consumption optimization model for the capacitated vehicle routing problem. Comput. Oper. Res. **39**, 1419–1431 (2012)
216. Xiao, Y., Zuo, X., Kaku, I., Zhou, S., Pan, X.: Development of energy consumption optimization model for the electric vehicle routing problem with time windows. J. Clean. Prod. **225**, 647–663 (2019)
217. Xue, J., Grift, T.E., Hansen, A.C.: Effect of biodiesel on engine performances and emissions. Renew. Sustain. Energy Rev. **15**, 1098–1116 (2011)
218. Yang, J., Sun, H.: Battery swap station location-routing problem with capacitated electric vehicles. Comput. Oper. Res. **55**, 217–232 (2015)
219. Yavuz, M.: An iterated beam search algorithm for the green vehicle routing problem. Networks. **69**, 317–328 (2017)
220. Yavuz, M., Çapar, İ.: Alternative-fuel vehicle adoption in service fleets: impact evaluation through optimization modeling. Transp. Sci. **51**, 480–493 (2017)
221. Yin, P.Y., Chuang, Y.L.: Adaptive memory artificial bee colony algorithm for green vehicle routing with cross-docking. Appl. Math. Model. **40**, 9302–9315 (2016)
222. Yu, Y., Tang, J., Li, J., Sun, W., Wang, J.: Reducing carbon emission of pickup and delivery using integrated scheduling. Transp. Res. Part D Transp. Environ. **47**, 237–250 (2016)
223. Yu, V.F., Perwira Redi, A.A.N., Hidayat, Y.A., Wibowo, O.J.: A simulated annealing heuristic for the hybrid vehicle routing problem. Appl. Soft Comput. **53**, 119–132 (2017)
224. Yu, Y., Wang, S., Wang, J., Huang, H.: A branch-and-price algorithm for the heterogeneous fleet green vehicle routing problem with time windows. Transp. Res. Part B. **122**, 511–527 (2019)
225. Yu, Y., Wu, Y., Wang, J.: Bi-objective green ride-sharing problem: model and exact method. Int. J. Prod. Econ. **208**, 472–482 (2019)
226. Yuan, X., Zhang, C., Hong, G., Huang, X., Li, L.: Method for evaluating the realworld driving energy consumptions of electric vehicles. Energy. **141**, 1955–1968 (2017)
227. Zheng, Y., Dong, Z.Y., Xu, Y., Meng, K., Zhao, J.H., Qiu, J.: Electric vehicle battery charging/swap stations in distribution systems: comparison study and optimal planning. IEEE Trans. Power Syst. **29**, 221–229 (2014)
228. Zhang, H., Hu, Z., Xu, Z., Song, Y.: Evaluation of achievable vehicle-to-grid capacity using aggregate PEV model. IEEE Trans. Power Syst. **32**, 784–79 (2017)
229. Zhang, D., Wang, X., Li, S., Ni, N., Zhang, Z.: Joint optimization of green vehicle scheduling and routing problem with time-varying speeds. PloS One. **13**, 1–20 (2018)
230. Zhang, S., Gajpal, Y., Appadoo, S.S., Abdulkader, M.M.S.: Electric vehicle routing problem with recharging stations for minimizing energy consumption. Int. J. Prod. Econ. **203**, 404–413 (2018)

231. Zhang, X., Yao, J., Liao, Z., Li, J.: The electric vehicle routing problem with soft time windows and recharging stations in the reverse logistics. In: International Conference on Management Science and Engineering Management, pp. 171–182. Springer, Cham (2018)
232. Zhang, S., Chen, M.Z., Zhang, W.Y.: A novel location-routing problem in electric vehicle transportation with stochastic demands. J. Clean. Prod. **221**, 567–581 (2019)
233. Zhang, S., Zhang, W., Gajpal, Y., Appadoo, S.S.: Ant colony algorithm for routing alternate fuel vehicles in multi-depot vehicle routing problem. In: Decision Science in Action: Theory and Applications of Modern Decision Analytic Optimisation, pp. 251–260. Springer, Singapore (2019)
234. Zhen, L., Xu, Z., Ma, C., Xiao, L.: Hybrid electric vehicle routing problem with mode selection. Int. J. Prod. Res. **58**(2), 562–576 (2020)
235. Zhenfeng, G., Yang, L., Xiaodan, J., Sheng, G.: The electric vehicle routing problem with time windows using genetic algorithm. In: IEEE 2nd Advanced Information Technology, Electronic and Automation Control Conference, pp. 635–639 (2017)
236. Zhou, B.-h., Tan, F.: Electric vehicle handling routing and battery swap station location optimisation for automotive assembly lines. Int. J. Comput. Integr. Manuf. **31**, 978–991 (2018)

# Chapter 2
# A Robust Optimization for a Home Healthcare Routing and Scheduling Problem Considering Greenhouse Gas Emissions and Stochastic Travel and Service Times

**Amir Mohammad Fathollahi-Fard , Abbas Ahmadi, and Behrooz Karimi**

**Abstract** Nowadays, there is a great deal of interest in employing home health care (HHC) services for elderly people in developed countries. An HHC system aims to consider a number of caregivers started from a pharmacy, serving a set of patients at their home with different home care services including housekeeping, nursing, and physiotherapist. Next, they analyze the biological samples from patients to update their health records in a laboratory. The logistics activities of this system need efficient optimization models and algorithms to find a valid plan for the caregivers addressing the HHC routing and scheduling problem (HHCRSP). To meet the demand of patients in a timely fashion, the caregiver has to deal with some stochastic parameters to carry out the valid plan to visit the patients. To achieve sustainability for HHC organizations, the gas dimensions of these logistic activities should be investigated. Although there are many papers in this research area from recent years, there is no study to consider the gas emissions as well as the stochastic travel and service times by using a robust optimization model. A bi-objective robust optimization model is developed for the first time and solved by a set of efficient metaheuristic algorithms. Having already been applied to similar problems, the Keshtel algorithm (KA) has been never employed in this research area. Therefore, another contribution of this study is to apply the KA and to compare it with well-known algorithms from the literature. Finally, an extensive comparison by different multi-objective optimization criteria confirms the performance of the proposed KA as well as the efficiency of the developed robust optimization approach.

A. M. Fathollahi-Fard · A. Ahmadi (✉) · B. Karimi
Department of Industrial Engineering and Management Systems, Amirkabir University
of Technology (Tehran Polytechnic), Tehran, Iran
e-mail: abbas.ahmadi@aut.ac.ir; b.karimi@aut.ac.ir

© Springer Nature Switzerland AG 2020    43
H. Derbel et al. (eds.), *Green Transportation and New Advances in Vehicle
Routing Problems*, https://doi.org/10.1007/978-3-030-45312-1_2

## 2.1  Introduction

The home health care (HHC) is related to the informal care for elderly people
at their home instead of hospitals or retirement homes [1]. The HHC services
are usually less expensive, more effective, and convenient than the formal care
[2]. Since the number of aging population is growing dramatically, the needs and
benefits from HHC services confirm their economic role to create more jobs such
as nurses, doctors, physiotherapists, and nutritionists [3]. The HHC industry has
become one of the important sectors of today's economy in developed countries
such as Germany, Austria, Sweden, Japan, and the USA [4].

The basic steps of an HHC organization's activities are described next. The first
step is to collect the patients' information from their electronic health records. The
second step is related to the decision-makers of the HHC organization. A valid plan
is needed to support the assignment of patients as well as routing and scheduling
of the caregivers by entirely taking into account the patients' data and the limited
resources. Finally, the caregivers execute this plan by driving a vehicle and visiting
the patients on their list. For each patient, the needed home care services and
medications are provided by the caregiver. After all visits, the biological samples
and treatments taken from the patients will be analyzed in a laboratory to update the
patients' health records.

The key elements to find a valid plan are the travel and service times. As
reported by HHC companies [5], according to the second step, the travel and service
times are deterministic. However, the caregivers have to deal with different types
of uncertainties for these important parameters. There are some common factors
such as weather and road conditions, driving skills, and unpredicted events for
unavailability of patients, which lead to uncertainty in travel and service times [6].
To control these stochastic parameters, among the first studies in this research area,
this study deploys a robust optimization to address an HHC routing and scheduling
problem (HHCRSP) by considering green dimensions of the logistic activities.

The HHCRSP includes several optimization decisions from the assignment of
patients to the closest pharmacy, scheduling, and routing decisions of caregivers
[7]. The transportation costs can affect these decisions and the goal in many studies
is to find a valid planning to reach the optimal cost [8]. In addition to the economic
factors of transportation, nowadays, sustainable transportation refers to any means
of transportation which is green and has a low impact on the environment [9].
The transportation sectors are generally responsible for a large portion of the
pollution for greenhouse gas (GHG) emissions worldwide [9]. Accordingly, this
study introduces a bi-objective HHCRSP considering GHG emissions as a separate
objective function.

The HHCRSP is known as an NP-hard problem [4, 5, 7, 8]. The literature shows a
great deal of interest in applying heuristics and metaheuristics to solve the HHCRSP.
The main reason is that current exact algorithms have poor performance when
addressing such complex problems within limited computation time. Therefore,
scholars always try to employ new algorithms in this research area. Although Kesh-

tel algorithm (KA) [10] as a powerful nature-inspired algorithm was successfully applied to many real-world optimization problems, it has never been applied in the area of HHCRSP. This study for the first time applies KA to a bi-objective HHCRSP considering GHG emissions and stochastic travel and service times and compares it with well-established algorithms from the literature.

Generally speaking, this paper highlights the following contributions as compared to existing papers from the literature.

1. A robust optimization model is developed to address a bi-objective HHCRSP for the first time.
2. This proposed HHCRSP contributes not only to stochastic travel and service times, but also gas emissions of logistics activities.
3. The Keshtel algorithm is applied to the proposed problem as compared with well-known algorithms based on different criteria.

The rest of this study is organized as follows: Section 2.2 reviews important and recent related studies to identify the research gaps. Section 2.3 establishes the proposed bi-objective HHCRSP for both deterministic and robust optimization versions. The solution approach is defined in Sect. 2.4 where the encoding plan is shown and the proposed KA is introduced. An extensive comparison based on different criteria is done in Sect. 2.5. Finally, the conclusion and future studies are discussed in Sect. 2.6.

## 2.2    Literature Review

The literature review of HHCRSP is still new and there are about seventy papers in Scopus from 1997 till now [5]. Based on this, one of the first studies can be referred to Begur et al. [11]. They applied a decision-making technique called spatial decision support system (SDSS) to model a simplified HHC through a case study in the USA. Nine years later, the first vehicle routing optimization was applied to the HHC by Bertels and Fahle [12] in 2006. They solved it by a heuristic algorithm combining the linear and constraint programming models and local search algorithms. In 2007, a particle swarm optimization algorithm (PSO) was applied successfully by Akjiratikarl et al. [13] to address an HHC scheduling problem in Ukraine. In 2009, Eveborn et al. [14] developed two mathematical models with the goal of optimization to improve both HHC routing and scheduling problems. Later in 2011, Trautsamwieser et al. [15] addressed an HHC scheduling problem under uncertainty. Based on an inspiration from a flood in Austria in 2002, a stochastic programming model was developed for the first time in this research area. Another HCC scheduling model was introduced by Rasmussen et al. [16] who applied a branch-and-price algorithm to solve it.

Since the logistic activities of HHC play a key role in the total cost, the HHC routing optimization shows a great deal of attention rather than HHC scheduling. In 2013, Liu et al. [17] proposed an integrated HHC routing with scheduling and time windows to develop an HHCRSP. Both pickup and delivery operations were considered for distributing the medications and collecting the biological samples. Another main innovation of their work was to introduce some heuristics based on the genetic algorithm (GA) and Tabu search (TS). In 2014, Liu et al. [18] introduced a hybrid metaheuristic based on TS with feasible and infeasible local searches to tackle an HHCRSP. In another research, Mankowska et al. [19] offered an HHCRSP considering multi-period and different types of home care services. Based on the availability of the caregivers, each of them can support two home care services. In 2015, Fikar and Hirsch [8] developed another HHCRSP with the possibility of caregivers' walking to visit the patients as well as the time window limitations. A case study in Austria was suggested to generate the test problems and two hybrid metaheuristics were introduced based on TS and simulated annealing (SA). In 2016, a bi-objective optimization model for the HHCRSP was firstly developed by Braekers et al. [20]. Their goal was to find an interaction between two minimization objectives, i.e., the total cost and the patients' inconvenience. To solve their model, they also used some simplified metaheuristics combining with dynamic programming. In 2017, Shi et al. [21] proposed a single objective and single depot HHCRSP for the first time by using a fuzzy demand function.

In 2018, Lin et al. [22] offered a new integrated HHCRSP by using a hybrid metaheuristic based on harmony search algorithm (HSA) and GA. They also added inheritance and immigrant schemes to this problem to consider the synchronization of the caregivers. Liu et al. [23] developed another bi-objective optimization model to consider medical team working in home care services. In addition to the transportation costs, they minimized the unemployment time of caregivers through group medical operations. In another paper, a memetic algorithm (MA) for the first time was applied by Decerle et al. [1] to a single objective multi-period HHCRSP. A Lagrangian relaxation-based algorithm was firstly applied to this problem by Fathollahi-Fard et al. [3]. They also considered a penalty function to better optimize the routes of caregivers also known as travel balancing. In another study, Cappanera et al. [24] for the first time proposed a robust optimization approach for an HHCRSP under demand uncertainty. In another bi-objective optimization model, Fathollahi-Fard et al. [25] considered the gas emissions and environmental pollution for an HHCRSP for the first time. They also utilized salp swarm algorithm (SSA) as a new metaheuristic and its hybridization with SA for the first time in this research area. Another version of HHCRSP under uncertainty was developed by Shi et al. [26] who applied bat algorithm (BA) and firefly algorithm (FA) to solve the problem and compared it with GA and SA.

Recently, a dynamic accepting rule for the patients scheduling to support the home care services based on the total cost and the availability of caregivers was proposed firstly by Demirbilek et al. [27]. Their main new supposition was that the patients arrived dynamically in each time period. Fathollahi-Fard et al. [7] developed three fast heuristics and a lower bound as well as a hybrid of variable neighborhood search (VNS) and SA to solve an HHCRSP. Lastly, Shi et al. [28] proposed a new robust optimization for the mentioned problem considering uncertain travel and service times. They applied two simplified metaheuristics based on SA and VNS to solve their proposed problem.

The literature review is summarized in Table 2.1. We classify the papers in terms of six criteria based on the number of objective functions, number of depots, number of time periods, and the outputs of model including the assignment of patients to pharmacies, routing, and scheduling of the caregivers. Another group of papers is classified by some attributes in the models such as the time windows, delivery time, synchronization, travel balancing, working time balancing, uncertainty approaches, and gas emissions. The last item is based on the solution algorithms. Based on these classifications, the following findings can be concluded:

1. Many studies in the area of HHCRSP have a single objective, involve only one depot and single period.
2. There is no study to consider all multi-depot, multi-period, and multi-objective HHCRSP simultaneously. Notably, a multi-period HHCRSP is able to cover the dynamic environment of this problem.
3. The gas emissions were only addressed in one study [25]. However, their model was deterministic and considered a single depot and period.
4. There are a few multi-objective HHCRSPs in the literature [4, 20, 23, 25].
5. Uncertainty modeling is still scarce and was considered by few researchers [21, 24, 26, 28].
6. Stochastic travel and service times were only assumed by Demirbilek et al. [27]. However, their model was single depot, single period, and single objective.
7. The suppositions of the time window, delivery time, travel balancing, and uncertainty of parameters and gas emissions are firstly contributed by this study.
8. The GA and PSO were successfully applied in this research area [13, 17, 21, 22, 28].
9. Already applied to other optimization problems, KA has been never applied to this research area.

To fill the research gaps, this study proposes a bi-objective HHCRSP considering GHG emissions and stochastic travel and service times. To tackle the uncertainties, a robust optimization model is developed. The proposed model is a type of multi-depot, multi-period, and multi-objective HHCRSP. In addition to the time window and delivery time characteristics, there are also different vehicles with various GHG emissions and travel balancing limitations similar to many recent studies [3, 7, 20, 23, 25]. Last but not least, a multi-objective version of KA is applied to solve the proposed problem. This algorithm is compared with two well-established algorithms, i.e., non-dominated sorting GA (NSGA-II) and multi-objective PSO (MOPSO).

**Table 2.1** Literature review

| Reference | Number of objectives | | Number of depots | | Number of periods | | Outputs of the model | | | Suppositions of the model | | | | | Uncertainty | Gas emissions | Solution algorithm |
|---|---|---|---|---|---|---|---|---|---|---|---|---|---|---|---|---|---|
| | Single objective | Multi-objective | Single depot | Multi-depot | Single period | Multi-period | Assignment of patients | Routing of caregivers | Scheduling of caregivers | Time windows | Delivery time | Synchronization | Travel balancing | Working time balancing | | | |
| Begur et al. [11] | ✓ | – | ✓ | – | ✓ | – | – | ✓ | – | – | – | – | – | – | – | – | SDSS |
| Bertels and Fahle [12] | ✓ | – | ✓ | – | ✓ | – | – | ✓ | – | | – | – | – | – | – | – | Hyper heuristic |
| Akjiratikarl et al. [13] | ✓ | – | ✓ | – | ✓ | – | – | – | ✓ | – | – | – | – | – | – | – | PSO |
| Eveborn et al. [14] | ✓ | – | ✓ | – | ✓ | – | ✓ | ✓ | ✓ | ✓ | – | – | – | – | – | – | Exact |
| Trautsamwieser et al. [15] | ✓ | – | ✓ | – | ✓ | – | – | ✓ | ✓ | – | ✓ | – | – | – | ✓ | – | VNS |
| Rasmussen et al. [16] | ✓ | – | ✓ | – | ✓ | – | – | ✓ | ✓ | ✓ | ✓ | – | – | – | – | – | Exact |
| Liu et al. [17] | ✓ | – | ✓ | – | ✓ | – | – | ✓ | ✓ | ✓ | ✓ | – | – | – | – | – | GA and TS |
| Liu et al. [18] | ✓ | – | ✓ | – | ✓ | – | – | ✓ | ✓ | ✓ | ✓ | – | – | – | – | – | Feasible rules for TS |
| Mankowska et al. [19] | ✓ | – | ✓ | – | ✓ | – | – | ✓ | ✓ | ✓ | ✓ | – | – | ✓ | – | – | Exact |
| Fikar and Hirsch [8] | ✓ | – | ✓ | – | – | ✓ | – | ✓ | ✓ | ✓ | ✓ | – | – | ✓ | – | – | TS |
| Braekers et al. [20] | – | ✓ | ✓ | – | – | ✓ | – | ✓ | ✓ | ✓ | ✓ | – | – | – | – | – | Dynamic metaheuristic |
| Shi et al. [21] | ✓ | – | ✓ | – | ✓ | – | – | ✓ | ✓ | ✓ | ✓ | – | ✓ | – | ✓ | – | Hybrid of GA and SA |

(continued)

| Author | | | | | | | | | | | | | | Method |
|---|---|---|---|---|---|---|---|---|---|---|---|---|---|---|
| Shi et al. [26] | √ | – | – | √ | – | √ | √ | √ | √ | – | – | √ | – | GA, SA, BA and FA |
| Cappanera et al.. [24] | √ | – | – | √ | – | √ | √ | √ | √ | – | – | √ | – | Exact |
| Lin et al. [22] | √ | – | √ | √ | √ | √ | √ | √ | √ | – | √ | – | – | Hybrid of HSA and GA |
| Liu et al.[23] | – | √ | √ | √ | √ | √ | √ | √ | √ | – | √ | – | – | Exact |
| Fathollahi-Fard et al. [25] | – | √ | – | √ | – | √ | √ | √ | √ | √ | – | √ | √ | Heuristics, SA and SSA |
| Decerle, et al. [1] | √ | – | – | √ | – | √ | √ | √ | √ | – | – | √ | – | MA |
| Fathollahi-Fard et al. [3] | √ | – | – | √ | – | √ | √ | √ | √ | – | – | √ | √ | Lagrangian relaxation |
| Bahadori-Chinibelagh et al. [29] | √ | – | – | √ | – | √ | √ | √ | √ | √ | – | – | – | Heuristics |
| Demirbilek et al. [26] | √ | – | √ | √ | √ | √ | – | √ | √ | – | √ | – | √ | Exact |
| Grenouilleau et al. [30] | √ | – | √ | √ | √ | √ | – | √ | √ | √ | √ | – | – | Heuristics |
| Decerle, et al. [4] | – | √ | – | √ | √ | √ | √ | √ | √ | – | – | √ | √ | Hybrid of MA and ACO |

(continued)

**Table 2.1** (continued)

| Reference | Number of objectives | | Number of depots | | Number of periods | | Outputs of the model | | | Suppositions of the model | | | | | | | Solution algorithm |
|---|---|---|---|---|---|---|---|---|---|---|---|---|---|---|---|---|---|
| | Single objective | Multi-objective | Single depot | Multi-depot | Single period | Multi-period | Assignment of patients | Routing of caregivers | Scheduling of caregivers | Time windows | Delivery time | Synchronization | Travel balancing | Working time balancing | Uncertainty | Gas emissions | |
| Fathollahi-Fard et al. [7] | ✓ | – | ✓ | – | ✓ | – | – | ✓ | ✓ | ✓ | ✓ | – | ✓ | – | – | – | Heuristics and hybrid of VNS and SA |
| Shi et al. [28] | ✓ | – | ✓ | – | ✓ | – | – | ✓ | ✓ | ✓ | ✓ | – | – | – | ✓ | – | VNS, TS and SA algorithms |
| This study | – | ✓ | – | ✓ | – | ✓ | ✓ | ✓ | ✓ | ✓ | ✓ | – | ✓ | – | ✓ | ✓ | MOKA, NSGA-II, MOPSO |

## 2.3 Problem Definition

In the problem, each pharmacy employs a group of caregivers to provide the homecare services and supplies the medications required by the patients. First of all, the patients are assigned to the nearest one among the existing pharmacies. It should be mentioned that the locations of pharmacies and laboratories are fixed.

After finding the nearest depot for each patient, in each time period, a caregiver starts from his/her pharmacy and after visiting the allocated patients goes to his/her laboratory to analyze the biological samples and to update the patients' health records. To illustrate a solution for the investigated problem, a graphical example is given in Fig. 2.1.

There are some important uncertainties with respect to travel and service times. They have a direct impact on the planning of HHCRSP. To handle these uncertainties, a robust optimization is developed. Another main contribution of this paper is to focus on carbon emissions by vehicles.

### 2.3.1 Assumptions

The main assumptions of the proposed HHCRSP are:

1. A robust optimization is developed for an HHCRSP under uncertainty.
2. The proposed problem is bi-objective, multi-depot, and multi-period HHCRSP.
3. In addition to the total cost, the gas emissions of the vehicles are minimized.
4. Each time period is estimated as 8 h.
5. There is no distinction between different types of medications.
6. There are many distributed patients and an equal number of pharmacies and laboratories.

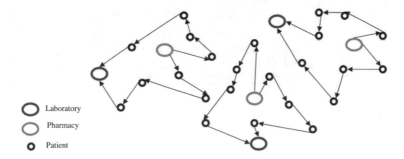

**Fig. 2.1** Description of the proposed multi-depot HHCRSP [7]

7. Each pharmacy and laboratory have a number of caregivers.
8. There is an allocation cost to assign the patients to pharmacies.
9. There are different types of vehicles with a specified capacity and transportation cost as well as amount of carbon emissions to distribute the required medications and to collect the biological samples.
10. The demand of each patient must be met and can be assigned to one pharmacy.
11. The capacity of each pharmacy and each laboratory in each location is limited.
12. The start point of each caregiver is his/her pharmacy. Similarly, the end point of the caregiver's route is his/her laboratory.
13. The caregivers supply the required medications of patients from the pharmacy; after visiting the patients, their biological samples should be assigned to the laboratories.
14. The stochastic parameters are the travel and service times including the time window based on the availability of patients at their earliest time and the latest time as well as the working time for each patient.
15. There is a limitation for distances traveled by the caregivers as a penalty function to do the travel balancing in the first objective.

### 2.3.2 Deterministic Model

To define the deterministic model as a mixed integer linear programming model, the following notations are considered. It should be noted that due to differences between uncertain parameters and deterministic ones, the tilde sign ($\sim$) has been utilized for uncertain input parameters.

| Indices | |
|---|---|
| $k$ | Index of vehicles, $k \in \{1, 2, \ldots, K\}$ |
| $i, j$ | Index of patients, $i, j \in \{1, 2, \ldots, M\}$ |
| $p$ | Index of pharmacies, $p \in \{1, 2, \ldots, P\}$ |
| $l$ | Index of laboratories, $l \in \{1, 2, \ldots, L\}$ |
| $n$ | Index of caregivers for each pharmacy, $n \in \{1, 2, \ldots, N_p\}$ |
| $t$ | Index of time periods, $t \in \{1, 2, \ldots, T\}$ |

| Parameters | |
|---|---|
| $D_{ij}^M$ | Distance between patients $i$ and $j$ |
| $D_{ip}^P$ | Distance between patient $i$ and pharmacy $p$ |
| $D_{pl}^L$ | Distance between pharmacy $p$ and laboratory $l$ |
| $CAPP_p$ | The capacity of pharmacy $p$ |
| $CAPL_l$ | The capacity of laboratory $l$ |
| $CS$ | The allocation cost per unit of distance |
| $TC_k$ | The transportation cost for each traveled distance regarding the vehicle $k$ |
| $CAP_k$ | The capacity of vehicle $k$ |
| $CER_k$ | $CO_2$ emissions rate for vehicle $k$ |
| $FCR_k$ | Rate of fuel consumption for vehicle $k$ |
| $GE_{ij}^k$ | Amount of $CO_2$ emissions generated by vehicle $k$ for traveling distance between nodes $i$ and $j$ ($GE_{ij}^k = CER_k \times FCR_k \times D_{ij}^M$) |
| $\tilde{W}_{it}$ | The service time to the patient $i$ in time period $t$ |
| $\tilde{E}_{it}$ | The earliest time of servicing to the patient $i$ in time period $t$ |
| $\tilde{L}_{it}$ | The latest time of servicing to the patient $i$ in time period $t$ |
| $\tilde{T}_{ijt}$ | The traveling time of patients $i$ to $j$ in time period $t$ |
| $PEN$ | Amount of penalty for overall distanced traveling between patients ($1 < PEN < 5$) |
| $BIG$ | A positive large number for constructing of constraints |
| $MDIS_{nkt}$ | Maximum desired traveling distances regarding each caregiver $n$ by employing vehicle $k$ in time period $t$ |
| $A_{it}$ | Demands of patient $i$ in time period $t$ |
| $B_{it}$ | Amount of biological samples received from patient $i$ after visiting by the caregiver in time period |

| Decision variables | |
|---|---|
| $X_{ijnpl}^{kt}$ | It gets 1 if the caregiver $n$ started from pharmacy $p$ to laboratory $l$ by applying vehicle $k$ visits the patients $i$ before $j$ in time period $t$; otherwise 0. |
| $Z_{ip}^P$ | It gets 1 if patient $i$ assigned to pharmacy $p$, otherwise 0. |
| $Z_{pl}^L$ | It gets 1 if pharmacy $p$ assigned to laboratory $l$, otherwise 0. |
| $S_{inpl}^t$ | Denotes the time at which caregiver $n$ started from pharmacy $p$ to laboratory $l$ begins to service the patient $i$ in time period $t$. |
| $O_{nkpl}^t$ | The overall traveled distances for caregiver $n$ from pharmacy $p$ to laboratory $l$ using vehicle $k$ in time period $t$. |

The proposed formulation is established as follows:

$$Z_1 = \min \sum_{i=1}^{M} \sum_{p=1}^{P} CSD_{ip}^{P} Z_{ip}^{P}$$

$$+ \sum_{p=1}^{P} \sum_{l=1}^{L} CSD_{pl}^{L} Z_{pl}^{L} \sum_{k=1}^{K} \sum_{n=1}^{N_p} \sum_{i=1}^{M} \sum_{j=1}^{M} \sum_{p=1}^{P} \sum_{l=1}^{L} \sum_{t=1}^{T} D_{ij}^{M} TC_k X_{ijnpl}^{kt}$$

$$+ \sum_{n=1}^{N_p} \sum_{k=1}^{K} \sum_{p=1}^{P} \sum_{l=1}^{L} \sum_{t=1}^{T} O_{nkpl}^{t} TC_k PEN) \tag{2.1}$$

$$Z_2 = \min \sum_{k=1}^{K} \sum_{n=1}^{N_p} \sum_{i=1}^{M} \sum_{p=1}^{P} \sum_{l=1}^{L} \sum_{t=1}^{T} \sum_{j=1}^{M} GE_{ij}^{k} X_{ijnpl}^{kt} \tag{2.2}$$

s.t.

$$\sum_{p=1}^{P} Z_{ip}^{P} = 1, \forall i \in M \tag{2.3}$$

$$\sum_{i=1}^{M} A_{it} \times Z_{ip}^{P} \leq CAPP_p, \forall p \in P, t \in T \tag{2.4}$$

$$\sum_{p=1}^{P} Z_{pl}^{L} = 1, \forall l \in L \tag{2.5}$$

$$\sum_{l=1}^{L} Z_{pl}^{L} = 1, \forall p \in P \tag{2.6}$$

$$\sum_{i=1}^{M} \sum_{j=1}^{M} \sum_{n=1}^{N_p} \sum_{k=1}^{K} \sum_{p=1}^{P} B_{it} X_{ijnpl}^{kt} \leq \sum_{p=1}^{P} CAPL_l Z_{pl}^{L}, \forall l \in L, t \in T \tag{2.7}$$

$$\sum_{n=1}^{N} \sum_{k=1}^{K} \sum_{j=1}^{M} \sum_{p=1}^{P} \sum_{l=1}^{L} X_{ijnpl}^{kt} = 1, \forall i \in M, t \in T \tag{2.8}$$

$$\sum_{i=1}^{M} A_i \times \sum_{j=1}^{M} X_{ijnpl}^{kt} \leq CAP_k, \forall k \in K, n \in N_p, p \in P, l \in L, t \in T \qquad (2.9)$$

$$\sum_{i=1}^{M} X_{ihnpl}^{kt} - \sum_{j=1}^{M} X_{hjnpl}^{kt} = 0, \forall h \in M, k \in K, n \in N_p, p \in P, l \in L, t \in T$$
$$(2.10)$$

$$S_{inpl}^t + \tilde{T}_{ijt} + \tilde{W}_{it} - BIG \times (1 - X_{ijnpl}^{kt}) \leq S_{jnpl}^t, \forall i, j \in M, k \in K, n \in N_p, p \in P, l \in L, t \in T$$
$$(2.11)$$

$$\tilde{E}_{it} \leq S_{inpl}^t \leq \tilde{L}_{it}, \forall i \in M, n \in N_p, p \in P, l \in L, t \in T \qquad (2.12)$$

$$O_{nkpl}^t \geq (\sum_{i=1}^{M} \sum_{j=1}^{M} D_{ij}^M \times X_{ijnpl}^{kt}) - MDIS_{nkt}, \forall p \in P, l \in L, n \in N_p, k \in K, t \in T$$
$$(2.13)$$

$$X_{ijnpl}^{kt} \leq Z_{pl}^L, \forall i, j \in M, t \in T, l \in L, p, n \in N_p \qquad (2.14)$$

$$S_{inpl}^t, O_{nkpl}^t \geq 0 \qquad (2.15)$$

$$X_{ijnpl}^{kt}, Z_{ip}^P, Z_{pl}^L \in \{0, 1\} \qquad (2.16)$$

Equation (2.1) reveals the first objective function ($Z_1$) remaining the total cost of the proposed home healthcare routing and scheduling problem. In this regard, the first and second terms represent the allocation cost. The patients are allocated to their pharmacies in the first item. The second term represents the assigning of pharmacies to their laboratories. Generally, we have used an allocation based on the distance to reduce the transportation cost for traveling. There is no doubt that if we choose the closest patients for each pharmacy, the transportation costs will be reduced. It is true that the caregivers do not travel from the laboratory and pharmacy. However, if these two points are closed to each other, we can reduce the transportation costs.

In the last two terms, the routing decisions of caregivers are considered. With regard to the traveled distance of caregivers by applying a vehicle type $k$, the transportation cost is optimized. Finally, the overall distances of caregivers traveling are calculated. Notably, the penalty value for the overall distance is based on the choice of HHC organizations due to the importance of transportation costs.

The second objective function ($Z_2$) is given in Eq. (2.2) to minimize the amount of carbon emissions generated by the vehicles. This factor is directly related to fuel consumption of the vehicle ($FCR_k$), amount of carbon emissions per liter of fuel ($CER_k$), and the travel distance between the two nodes ($D_{ij}^M$).

Constraint (2.3) defines that each patient should be assigned to only one pharmacy. Constraint (2.4) ensures that the required medications of each patient to its pharmacy should be met. Constraint (2.5) makes sure that each laboratory should be assigned to one pharmacy only. Constraint (2.6) shows that each pharmacy should be assigned to only one laboratory. Constraint (2.7) specifies that the capacity of a laboratory should handle the whole biological samples during the visiting of patients. Constraint (2.8) indicates that the patients should be visited once only. Constraint (2.9) guarantees that the capacity of selected vehicles should be greater than the summations of the transported medications to the patients. Constraint (2.10) states that after visiting a patient, the caregiver should leave this patient. Regarding the limitations of time window, constraint (2.11) explores that the caregiver cannot arrive at patient $j$ before $S_{inpl}^{t} + \tilde{T}_{ijt} + \tilde{W}_{it}$. It specifies that each patient has a working time of $\tilde{W}_{it}$. Similarly, the patients traveling from $i$ to $j$ is considered by $\tilde{T}_{ijt}$. In this regard, $BIG$ is a large scalar number to construct this constraint. Constraint (2.12) shows that each patient has a time window in each time period. Constraint (2.13) aims to compute the extra traveling distance of caregivers limited by a maximum desired distance. Constraint (2.14) shows the relation of assignment and routing decisions. Constraints (2.15) and (2.15) finally impose the none-negativity and binary restrictions on the decision variables.

### 2.3.3   Robust Optimization Version

The deterministic version of HHCRSP ignores the uncertainty in the travel and service times. To make this model more reasonable and suitable for these uncertainties, a robust optimization approach firstly proposed by Ben-Tal and Nemirovski [31] is applied to this HHCRSP.

In this method, we firstly define two main sources of uncertainty as follows:

| $U s_k$ | Uncertainty set of service time for each vehicle k |
|---------|----------------------------------------------------|
| $U t_k$ | Uncertainty set of travel time for each vehicle k  |

The uncertainty set $U s_k$ refers to the working time of HHC services $\tilde{W}_{it}$. Following formula is defined:

$$U s_k = \{ \tilde{W}_{it} \in R^{\left| M^{N_p} \right|} \mid \tilde{W}_{it} = W_{it}^{ave} + \alpha_{it} W_{it}^{err}, \sum_{i=1}^{M^{N_p}} |\alpha_{it}| \leq \lambda s_k, |\alpha_{it}| \leq 1,$$

$$\lambda s_k = \left[ \theta s \left| M^{N_p} \right| \right], i \in M, t \in T \} \tag{2.17}$$

Equation (2.17) shows the uncertainty set of the service time for patient $i$ and caregiver $n$. In this regard, $M^{N_p}$ refers to the set of patients in a route served by caregiver $n$ and vehicle $k$. $\tilde{W}_{it}$ shows the stochastic amount of service time for patient $i$ and period time $t$. $W_{it}^{ave}$ represents the nominal value of this parameter as an average. $\alpha_{it}$ is an auxiliary number. $W_{it}^{err}$ is the variance from nominal value.

$\lambda s_k$ is the budget of uncertainty and its amount controls the level of uncertainty for the service time. Lastly, $\theta s \in [0, 1]$ is a coefficient of the uncertainty for the service time.

$$Ut_k = \{\tilde{T}_{ijt} \in R^{\left|M^{N_p}\right|} \big| \tilde{T}_{ijt} = T_{ijt}^{ave} + \beta_{ijt} T_{ijt}^{err}, \sum_{i=1}^{M^{N_p}} \sum_{j=1}^{M^{N_p}} \left|\beta_{ijt}\right| \leq \lambda t_k,$$

$$\left|\beta_{ijt}\right| \leq 1, \lambda t_k = \left[\theta t \left|M^{N_p}\right|\right], i, j \in M, t \in T\} \tag{2.18}$$

In addition to this, Eq. (2.18) represents the uncertainty set $Ut_k$ which refers to the travel time $\tilde{T}_{ijt}$ on the arc of $(i, j)$ in each time period $t$. This equation is similar to Eq. (2.17). $\tilde{T}_{ijt}$ shows the stochastic amount of the travel time from patient $i$ to $j$ and period time $t$. $T_{ijt}^{ave}$ represents the nominal value of this parameter as its average. $\beta_{ijt}$ is an auxiliary number. $T_{ijt}^{err}$ is the variance from nominal value for the travel time. $\lambda t_k$ is the budget of uncertainty to control the level of uncertainty for the travel time. Lastly, $\theta t \in [0, 1]$ is a coefficient of the uncertainty for the travel time.

Uncertainty can affect the feasibility of the solution. The optimal solution may change with respect to each realization of uncertain parameters. Therefore, a robust solution is better than any other solution because it is better in terms of minimizing a certain objective such as maximum regret. Thus, the deterministic model given in Eqs. (2.1)–(2.16) is transformed into a robust optimization version as follows:

$$Z_1 = \min_{Robust} \left( \sum_{i=1}^{M} \sum_{p=1}^{P} CS \times D_{ip}^{P} \times Z_{ip}^{P} + \sum_{p=1}^{P} \sum_{l=1}^{L} CS \times D_{pl}^{L} \times Z_{pl}^{L} \right.$$

$$+ \sum_{k=1}^{K} \sum_{n=1}^{N_p} \sum_{i=1}^{M} \sum_{j=1}^{M} \sum_{p=1}^{P} \sum_{l=1}^{L} \sum_{t=1}^{T} D_{ij}^{M} \times TC_k \times X_{ijnpl}^{kt}(Us_k, Ut_k)$$

$$\left. + \sum_{n=1}^{N_p} \sum_{k=1}^{K} \sum_{p=1}^{P} \sum_{l=1}^{L} \sum_{t=1}^{T} O_{nkpl}^{t}(Us_k, Ut_k) \times TC_k \times PEN \right) \tag{2.19}$$

$$Z_2 = \min_{Robust} \left( \sum_{k=1}^{K} \sum_{n=1}^{N_p} \sum_{i=1}^{M} \sum_{p=1}^{P} \sum_{l=1}^{L} \sum_{t=1}^{T} \sum_{j=1}^{M} GE_{ij}^{k} \times X_{ijnpl}^{kt}(Us_k, Ut_k) \right) \tag{2.20}$$

s.t.

$$\text{Eqs. } (2.3)\text{--}(2.10)* \tag{2.21}$$

$$S^t_{inpl}(Us_k, Ut_k) + T_{ijt} + W_{it} - BIG \times (1 - X^{kt}_{ijnpl}(Us_k, Ut_k)) \leq S^t_{jnpl}(Us_k, Ut_k)$$
$$, \forall i, j \in M, k \in K, n \in N_p, p \in P, l \in L$$
$$\tag{2.22}$$

$$E_{it} \leq S^t_{inpl}(Us_k, Ut_k) \leq L_{it}, \forall i \in M, n \in N_p, p \in P, l \in L, t \in T \tag{2.23}$$

$$\text{Eqs. } (2.13)\text{--}(2.16) \tag{2.24}$$

## 2.4 Proposed Solution

Here, the proposed solution algorithm is developed to address the underlying problem. First of all, an encoding plan is considered to address the proposed problem. Next, the details of the proposed multi-objective version of KA algorithm are given. We compare the proposed algorithm with two well-established metaheuristics from the literature. The NSGA-II and MOPSO are adopted to have this comparison. The details can be found in several recent studies [32, 33].

### 2.4.1 Encoding Plan

An encoding representation is needed to generate feasible solutions to cover the constraints of the model. One of the well-known techniques for the solution scheme is the random key [34]. This strategy uses two stages and saves time to search. It needs no repair to generate a feasible solution as its main benefit.

At first, the allocation variables including assigning of the patients to pharmacies ($Z^P_{ip}$) as well as the assignment of pharmacies to laboratories ($Z^L_{pl}$) are generated in Figs. 2.2 and 2.3, respectively. As indicated from Fig. 2.2, ten patients are allocated to three pharmacies by random numbers. For example, the patients including $m_2$, $m_4$, $m_5$, and $m_6$ are assigned to pharmacy $P_1$.

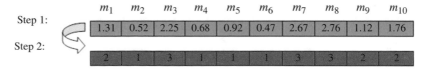

**Fig. 2.2** The allocation of patients to pharmacies

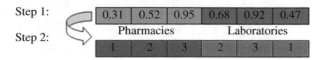

Fig. 2.3 The allocation of pharmacies and laboratories

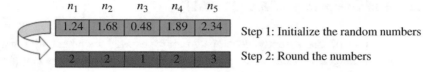

Fig. 2.4 The allocation type of vehicles to caregivers

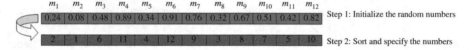

Fig. 2.5 The assignment of patients to each route of caregiver

Figure 2.3 represents the allocation of pharmacies to laboratories. The pharmacies and laboratories are allocated one by one. As can be seen, pharmacies $P_1$, $P_2$, and $P_3$ are assigned to laboratories $L_2$, $L_3$, and $L_1$, respectively.

To start the planning of caregivers, first of all, the used vehicles for each caregiver should be selected. The considered procedure is an array by a length of $N$. The uniform distribution of $U(0, K)$ is conducted to generate the first step of the utilized random key technique, where $K$ is the number of vehicle types. In the second step by rounding the generated numbers, the type of vehicles for each caregiver is clarified. These procedures are shown in Fig. 2.4. As can be envisaged from this figure, for caregiver $n_1$, $n_2$, and $n_4$, the selected vehicle is the second type. As such, regarding the caregivers $n_3$ and $n_5$, the first and third types of vehicles are chosen, respectively.

To choose the patients' route, an array with length of assigned patients to this pharmacy (in this example, it equals to 12) is distributed by $U(0, 1)$. Figure 2.5 reveals a simple instance with the used arrays for the generated matrix of adopted methodology. In this case, there are 12 patients. The second step is to sort these numbers to specify the routes. Notably, these routes should meet the maximum desired total distance of traveling ($MDIS_{nkt}$), the capacity of the used vehicle ($CAP_k$). The time window's limitation is considered as a penalty value based on similar works in this research area [3, 7, 21, 25, 26, 28]. Note that this encoding plan is repeated for each time period. As a result, the routes adopted from this example are numerically as follows:

$$n_1 = \{m_2 \rightarrow m_1 \rightarrow m_8\}$$

,

$$n_2 = \{m_5 \rightarrow m_{11} \rightarrow m_3\}$$

,

$$n_3 = \{m_{10} \rightarrow m_9 \rightarrow m_7 \rightarrow m_{12}\}$$

,

$$n_4 = \{m_4 \rightarrow m_6\}$$

### 2.4.2  Multi-Objective of KA (MOKA)

The KA as one of the recently developed nature-inspired metaheuristics was proposed by Hajiaghaei-Keshteli and Aminnayeri [10] in order to solve complex optimization problems. Already applied to several optimization problems such as distribution networks [33], production scheduling [35], closed-loop supply chains [36, 37], truck scheduling [38], the KA has been never applied to HHC optimization models. This study for the first time introduces this algorithm in this research area.

The Keshtel is the name of a bird. This bird shows a very interesting behavior after finding the food. The lucky Keshtels find better food in the lake [33]. Then, the Keshtels being in their neighborhood are attracted toward them and all at once start to swirl around the food source. During swirling, if a Keshtel finds a better food source, it will be identified as a lucky Keshtel. Moreover, several Keshtels move toward intact spots of the lake in order to find other foods [35]. During this movement, they consider the position of two other Keshtels. In the lake, there are also some other Keshtels that do not find any food. They leave the lake and are replaced with newcomer Keshtels. These facts taken from the nature are the main inspiration of KA.

Generally, this algorithm starts with initial random solutions. This population is divided into three groups. The first group is the best Keshtels and the last one is the worst Keshtels, conversely. The first group maintains to exploit the feasible search space. The second one tries to shake the local search and the third one improves the exploration properties. This study proposes a multi-objective of KA known as MOKA. After performing the procedures of KA, the next generation is updated by dominance technique [33, 37, 39]. The main difference of this algorithm from its original idea is the comparison between two Keshtels. If there is no improvement in the objective functions of Keshtels, the Pareto-optimal set is updated. Otherwise, new Keshtel will replace the current one. In the following, the main steps of the proposed MOKA are addressed.

#### 2.4.2.1  Primary Procedure

As mentioned earlier, population members are divided into three sections. Let $N$ denote the set of population members, then:

$$N = N_1 \cup N_2 \cup N_3 \tag{2.25}$$

where $N_1$ includes a number of population member bearing a better value for fitness function compared to the rest of members (lucky Keshtels), $N_2$ includes a number of population member bearing the worst value for fitness function compared to the rest of members, and $N_3$ includes population members which do not exist in $N_1$ and $N_2$ sets.

### 2.4.2.2 Attraction and Swirling

If a lucky Keshtel finds a food source in the lake, another Keshtel will be attracted toward it and swirl around the food source with a specific radius. Meanwhile, if it discovers a better food source (it means that this solution can dominate the lucky one), it is identified as a lucky Keshtel. Otherwise, after each swirl, it reduces the radius. This swirl continues until the food source is finished. During the swirling process, if the neighbor cannot dominate this solution, the Pareto set should be updated and the process would stop. This procedure is presented in Fig. 2.6.

### 2.4.2.3 Replace the Members of $N_2$ Set with New Ones

Not having been able to find any food, Keshtels leave the lake and new Keshtels hoping to find food come to the lake. Therefore, the members of set $N_2$ bearing worse value of objective function than that of other members are omitted and new members are produced randomly and replaced.

### 2.4.2.4 Move the Member of $N_3$ Set

Each member of $N_3$ set changes its position toward virgin spots in terms of other two members' position. Let $Y_i$ be a member of $N_3$ set. It changes its position as follows:

$$v_i = \lambda_1 \times Y_j + (1 - \lambda_1) \times Y_t \qquad (2.26)$$

$$Y_i = \lambda_2 \times Y_i + (1 - \lambda_2) \times v_i \qquad (2.27)$$

**Fig. 2.6** Attraction and swirling process [10]

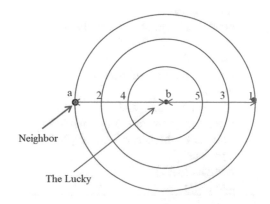

where $Y_j$ and $Y_t$ are two members selected randomly from $N_3$ set and are different from $Y_i$. $\lambda_1$ and $\lambda_2$ are random numbers selected from the uniform distribution in [0,1].

### 2.4.2.5 Stopping Condition

Aforementioned steps are continued and repeated until the stop condition such as the time interval or the maximum number of iteration would be satisfied. More details about its implementation are addressed by a pseudo-code given by Algorithm 1.

---

**Algorithm 1:** Pseudo-code of MOKA

---

1  Initialize Keshtels population.
   Calculate the fitness and sort them in three types: $N_1$, $N_2$ and $N_3$
   Generate the Pareto frontiers of solutions.
   By calculating the crowding distance in the best Pareto frontier, select the best Keshtel.
   $T_1$=clock;
   *while* ($t<$ maximum time of simulation)
   *for* each $N_1$
   Calculate the distance between this lucky Keshtel and all Keshtels.
   Select the closest neighbor.
   S=0;
   *while* (S< maximum number of swirling)
   Do the swirling.
   *if* the fitness (at least, one of objective functions has been improved) of this new position is better than prior
   Update this lucky Keshtel.
   *break*
   *endif*
   S=S+1
   *endwhile*
   *endfor*
   *for* each $N_2$
   Move the Keshtel between the two Keshtels based on Eqs. (2.26)–(2.27).
   *endfor*
   *for* each $N_3$
   Create a random solution.
   *endfor*
   Merge the $N_1$, $N_2$ and $N_3$ .
   Update the new Pareto-optimal solutions.
   Sort the Keshtels and form $N_1$, $N_2$ and $N_3$ for next iteration.
   $T_2$=clock;
   $t$=etime($T_2$,$T_1$);
   *end while*

---

## 2.5    Experimental Results

In this section, firstly, the test problems are determined to generate different problem complexities. Next, the algorithms are tuned. Based on the calibrated metaheuristics, an extensive comparison based on different criteria is done. The high-efficiency of the proposed MOKA and the performance of the proposed model are concluded. Note that all experiments have been done on an INTEL Core 2 CPU with a 2.4 GHz processor and 2 GB of RAM.

### 2.5.1    Instances

To address the model by MOKA, NSGA-II, and MOPSO, some benchmark tests from recent papers [3, 7, 25] are taken into consideration. The twelve instances are divided into three levels including small, medium, and large as reported in Table 2.2. Note that the range of parameters is based on the benchmarks [3, 7, 25].

### 2.5.2    Tuning

Since the employed metaheuristics, i.e., NSGA-II, MOPSO, and MOKA have some controlling parameters, it is essential to tune them satisfactorily. If the algorithms have not calibrated very well, the comparison would be unbiased [40, 41]. In this regard, this study utilizes a response surface method (RSM) [42] as one of the successful calibration methods. Notably, to evaluate the algorithms, multi-objective assessment metrics including the number of Pareto solutions (NPS) [39], spread of non-dominance solutions (SNS) [33], quality metric (QM) [37], and hypervolume (HV) [43] are utilized to evaluate the proposed HHCRSP.

In the RSM, specific factors are related to each of the metaheuristic input parameters. Each factor as a parameter $(x_i)$ is measured at two levels, coded as $-1$ to 1, relative to the low $(x_l)$ and high $(x_h)$ levels given by the selected range. Hence, the independent variables $(z_i)$ relating to each factor are generated by

$$z_i = \frac{x_i - (x_h + x_l)/2}{(x_h - x_l)/2}, i = \{1, 2, \ldots, K\} \tag{2.28}$$

where $K$ is the number of variables. To calculate the response of the independent variables, a polynomial response surface function $(y)$ is presented by

$$y = \beta_0 + \sum_{j=1}^{k} \beta_j z_j + \sum_{j=1}^{k} \sum_{i<j}^{k} \beta_{ij} z_i z_j + \sum_{j=1}^{k} \beta_{jj} z_{jj}^2 + \varepsilon \tag{2.29}$$

**Table 2.2** The size of instances

| Classification | Instance | Number of laboratories ($L$) and pharmacies ($P$) | Number of caregivers ($N_p$) | Types of vehicles ($K$) | Number of patients ($M$) | Number of periods ($T$) |
|---|---|---|---|---|---|---|
| Small | SP1 | 2 | 2 | 2 | 10 | 2 |
| | SP2 | 2 | 3 | 2 | 25 | 4 |
| | SP3 | 4 | 4 | 3 | 40 | 6 |
| | SP4 | 6 | 6 | 3 | 65 | 8 |
| Medium | MP5 | 8 | 8 | 3 | 80 | 14 |
| | MP6 | 9 | 8 | 4 | 85 | 18 |
| | MP7 | 9 | 9 | 5 | 95 | 24 |
| | MP8 | 10 | 10 | 5 | 100 | 28 |
| Large | LP9 | 12 | 12 | 6 | 120 | 32 |
| | LP10 | 14 | 15 | 6 | 150 | 36 |
| | LP11 | 16 | 16 | 7 | 160 | 40 |
| | LP12 | 18 | 20 | 8 | 200 | 42 |

**Table 2.3**  Metaheuristics and candidate factors

| Algorithm | Factors and their levels | | | | | Number of experiments; total number $= (n_f, n_{ax}, n_{cp})$ |
|---|---|---|---|---|---|---|
| MOKA | Maxit | nPop | N1 | N2 | Smax | $44 = (2^5, 6, 6)$ |
| | (300, 500) | (100, 200) | (0.1, 0.3) | (0.3, 0.6) | (2, 5) | |
| NSGA-II | Maxit | nPop | Pc | Pm | | $30 = (2^4, 8, 6)$ |
| | (200, 400) | (100, 200) | (0.5, 0.8) | (0.02, 0.1) | | |
| MOPSO | Maxit | nPop | C1 | C2 | W | $44 = (2^5, 6, 6)$ |
| | (300, 500) | (100, 200) | (1.5 1.25) | (1.5 1.25) | (0.8, 0.99) | |

where $\beta_0$ and $\beta_j$, $\beta_{ij}$, and $\beta_{jj}$ are the constants of the linear coefficient, the interaction coefficient ($\beta_{ij}$), and the quadratic coefficient ($\beta_{jj}$), respectively.

To start the RSM, the employed metaheuristics are given in Table 2.3, along with their factors based on their range. As such, the total number of experiments is measured by $n_f = 2^k$ as a fraction of normal treatments, $n_{ax} = 2k$ is the number of axial points, and $n_{cp}$ is the number of central points.

A utility function is applied to assess the metrics of each Pareto-optimal set and optimize the multiple responses of the RSM, as computed by

$$d_i (y_i) = (\frac{h_i - y_i}{h_i - l_i})^s . l_i < y_i < h_i \qquad (2.30)$$

where the multiple response $y_i$ has been transformed into the measurement of the utility function ($d_i$). $l_i$ and $h_i$ are the lower and upper bounds of response variables, respectively. The emphasis on the amount of utility function is calculated by $s$. Less emphasis on the assessment metrics equates with less importance. Accordingly, the amount of $s$ is 1, 1, 2, and 3, to reflect the relative importance of the evaluation metrics NPS, SNS, QM, and HV, respectively. The desirability of the algorithm in terms of the number of utility functions for all applied assessment metrics is computed by

$$D = \sqrt[m]{d_1(y_1)d_2(y_2)\ldots d_m(y_m)} \qquad (2.31)$$

where $m$ is the number of evaluation metrics. In this regard, $D$ is the total desirability of the algorithm. It is evident that the higher the value of $D$, the more favorable is the algorithm.

Table 2.4 presents the approximate values of the tuned parameters, the $R$-squared ($R^2$) of assessment metrics, and the total desirability ($D$). Notably, $R^2$ is a statistical measure of how closely the outputs fit to the regression line. The range of $R^2$ is always between 0 and 100%. Similar to the desirability, the higher the value of $R^2$, the more favorable is the algorithm.

**Table 2.4** Calibrated parameter values for each algorithm, their respective R-squared ($R^2$) and desirability (D)

| Optimizer | Tuned parameters | $R^2$ (%) | | | | D |
|---|---|---|---|---|---|---|
| | | NPS | QM | SNS | HV | |
| NSGA-II | *Maxit = 380; nPop = 140; Pc=0.8; Pm = 0.6;* | 54 | 72 | 60 | 58 | 0.6634 |
| MOKA | *Maxit = 410; nPop = 180; N1 = 0.25; N2 = 0.45; Smax = 3;* | 58 | 86 | 62 | 66 | 0.7238 |
| MOPSO | *Maxit = 350; nPop = 120; C1 = 2.1; C2 = 2.1; W = 0.92* | 54 | 74 | 68 | 72 | 0.7428 |

## *2.5.3 Comparison*

To start the comparison between MOKA and two well-known algorithms, first of all, their results must be checked by an exact solver. Since the proposed mathematical model is a type of multi-objective optimization model, the Epsilon Constraint method proposed by Haimes et al. [44] is applied to assess the non-dominated solutions of the metaheuristics. The structure of the EC is modeled by only one objective and the other objectives are considered as the constraints [37]. The proposed deterministic model is considered by the following formula:

$$\min Z_1 \qquad (2.32)$$

$$\text{s.t.}$$
$$\text{Eqs. (3) } - (16)$$
$$Z_2 \le \varepsilon$$
$$Z_2^{min} \le \varepsilon \le Z_2^{max}$$

With regard to the above formula, the optimal value for the first objective function ($Z_1$) is found. Next, the positive ($Z_2^{min}$) and negative ($Z_2^{max}$) idea solutions for the second objective function should be reached. In addition to these solutions, the allowable bound ($\varepsilon$) is also considered as the average point (($Z_2^{min} + Z_2^{max}$)/2). Finally, three non-dominated solutions are obtained by checking all the generated solutions.

Table 2.5 reveals the Pareto-optimal solutions in a small test problem, i.e., SP1 for both exact and metaheuristic algorithms. To make the comparison easier, the algorithms' solutions are sorted as given in the table. For validations, the non-dominated solutions of each metaheuristic by considering the non-dominated solutions of EC are highlighted. In this regard, the modified number of Pareto solution (MNPS) and its successful percentage, i.e., $\frac{MNPS}{NPS}$ have been calculated. Accordingly, their results are provided in Table 2.6.

**Table 2.5** Algorithms' Pareto solutions resulted in the test problem SP1

| Number | EC | | NSGA-II | | MOPSO | | MOKA | |
|---|---|---|---|---|---|---|---|---|
| | $Z_1$ | $Z_2$ | $Z_1$ | $Z_2$ | $Z_1$ | $Z_2$ | $Z_1$ | $Z_2$ |
| 1 | **10,001** | 2695 | **10,253** | 2720 | **10,146** | 2864 | **10,067** | 2791 |
| 2 | 11,432 | 2552 | 10,710 | 2714 | 10,310 | 2810 | 11,615 | 2766 |
| 3 | 12,970 | **2509** | 10,836 | 2712 | 10395 | 2786 | 12,018 | 2718 |
| 4 | – | – | 117,86 | 2696 | 10,606 | 2714 | 12,151 | 2650 |
| 5 | – | – | 11,913 | 2675 | 11,415 | 2650 | 12,237 | 2589 |
| 6 | – | – | 12,919 | **2672** | 11,931 | 2586 | 12,655 | 2536 |
| 7 | – | – | – | – | 12170 | **2523** | 12,662 | **2510** |

The best values are shown in bold

**Table 2.6** Validation of the applied metaheuristics

| Test problem | NSGA-II | | MOPSO | | MOKA | |
|---|---|---|---|---|---|---|
| | MNPS | MNPS/ NPS | MNPS | MNPS/ NPS | MNPS | MNPS/ NPS |
| SP1 | 0 | 0 | 2 | 0.33 | 2 | 0.28 |
| SP2 | 1 | 0.12 | 2 | 0.25 | 3 | 0.33 |
| SP3 | 2 | 0.28 | 1 | 0.11 | 3 | 0.3 |
| SP4 | 2 | 0.22 | 3 | 0.33 | 4 | 0.36 |
| MP5 | 3 | 0.3 | 2 | 0.22 | 3 | 0.3 |
| MP6 | 2 | 0.22 | 2 | 0.2 | 4 | 0.5 |
| Average | | 0.19 | | 0.24 | | **0.34** |

The best values are shown in bold

Having a conclusion about the validation of the applied metaheuristics, NSGA-II shows a weak performance, while MOPSO and MOKA reveal a high performance (Table 2.5). Based on the results of the test problems, i.e., SP1 to MP6, the strength of the MOKA is demonstrated in Table 2.6, where the average number of Pareto non-dominated solutions relative to the EC method is shown to be 0.34, and it is significantly higher than those of NSGA-II (2.19) and MOPSO (2.24).

The process time of the algorithms is the second criterion to compare the algorithms. The behavior of three selected metaheuristics is depicted in Fig. 2.7. The results obviously show that MOKA consumes more time as compared to NSGA-II and MOPSO for all test problems. The MOPSO has a neck-and-neck competition with NSGA-II and generally is the best algorithm in this item.

The results of the algorithms based on the assessment metrics of Pareto-optimal solutions, i.e., NPS, SNS, QM, and HV are provided in Table 2.7. To analyze them statistically, the results are transformed into a well-known metric called the relative deviation index (RDI) based on the following formula:

$$RDI = \frac{|ID_{sol} - AL_{sol}|}{MAX_{sol} - MIN_{sol}} \tag{2.33}$$

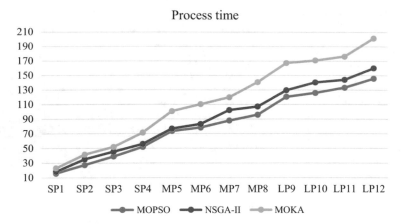

**Fig. 2.7** Process time of algorithms

where $AL_{sol}$ is the value given by the assessment metric to assess the Pareto-optimal sets. As such, $MAX_{sol}$ and $MIN_{sol}$ are the maximum and minimum possible values for the assessment metrics, respectively. Note that based on the nature of the assessment metric, $ID_{sol}$ would be one of $MAX_{sol}$ or $MIN_{sol}$. The lower the value of RDI, the better the performance of the algorithm is. In order to determine the efficiency of the five applied metaheuristics in comparison with each other, interval plots at the 95% statistical confidence level are provided in Fig. 2.8. These plots are based on the RDI metric as well.

According to Fig. 2.8, there is a set of clear differences between the performances of algorithms. Based on the NPS metric (Fig. 2.8a), MOPSO is the worst algorithm and conversely, MOKA is the best one. NSGA-II is slightly better than MOPSO. With regard to the SNS metric (Fig. 2.8b), NSGA-II and MOPSO are the same as the worst algorithms. Conversely, MOKA is the best algorithm in this item. Regarding the QM metric (Fig. 2.8c), the proposed MOKA is completely better than other algorithms. However, both MOPSO and NSGA-II have a same performance in this metric. Based on the HV metric (Fig. 2.8d), the algorithms have some similarities. The best algorithm is MOKA. Conversely, NSGA-II is slightly the worst one in this item.

In conclusion, the proposed MOKA reveals a high performance to solve the proposed HHCRSP in comparison with NSGA-II and MOPSO based on different criteria.

**Table 2.7** Results of assessment metrics for each metaheuristic

| Metaheuristic | | SP1 | SP2 | SP3 | SP4 | MP5 | MP6 | MP7 | MP8 | LP9 | LP10 | LP11 | LP12 |
|---|---|---|---|---|---|---|---|---|---|---|---|---|---|
| NSGA-II | NPS | 7 | 9 | 7 | 9 | 10 | 9 | 10 | 11 | 12 | 10 | 11 | 9 |
| | SNS | 47183 | 47199 | 44243 | 34634 | 72852 | 47079 | 41171 | 34479 | 30755 | 47890 | 61801 | 34218 |
| | QM | 0.15 | 0.12 | 0.16 | 0.18 | 0.12 | 0.2 | 0.15 | 0.1 | 0.13 | 0.08 | 0.09 | 0.18 |
| | HV | 6.06E+09 | 3.24E+09 | 3.52E+09 | 6.17E+09 | 4.42E+09 | 7.91E+09 | 5.86E+09 | 1.63E+09 | 5.99E+09 | 5.21E+09 | 7.73E+09 | 5.47E+09 |
| MOPSO | NPS | 7 | 9 | 10 | 9 | 9 | 10 | 8 | 9 | 11 | 10 | 9 | 6 |
| | SNS | 30727 | 61654 | 35015 | 35687 | 54184 | 33136 | 58925 | 71461 | 47700 | 35003 | 38369 | 43522 |
| | QM | 0.07 | 0.06 | 0.09 | 0.16 | 0.13 | 0.14 | 0.17 | 0.19 | 0.18 | 0.06 | 0.09 | 0.14 |
| | HV | 9.24E+09 | 6.53E+09 | 9.26E+09 | 4.26E+09 | 6.99E+09 | 1.51E+09 | 4.47E+09 | 2.09E+09 | 5.22E+09 | 6.78E+09 | 1.34E+09 | 4.4E+09 |
| MOKA | NPS | 8 | 10 | 11 | 10 | 11 | 9 | 10 | 9 | 11 | 12 | 10 | 9 |
| | SNS | 40218 | 46470 | 75252 | 41736 | 75354 | 76112 | 48919 | 77122 | 48167 | 66643 | 42386 | 50562 |
| | QM | 0.78 | 0.82 | 0.75 | 0.66 | 0.75 | 0.66 | 0.68 | 0.71 | 0.69 | 0.86 | 0.82 | 0.68 |
| | HV | 8.26E+09 | 4.9E+09 | 8.54E+09 | 5.4E+09 | 7.04E+09 | 4.11E+09 | 7.54E+09 | 8.24E+09 | 7.97E+09 | 6.71E+09 | 5.15E+09 | 7.22E+09 |

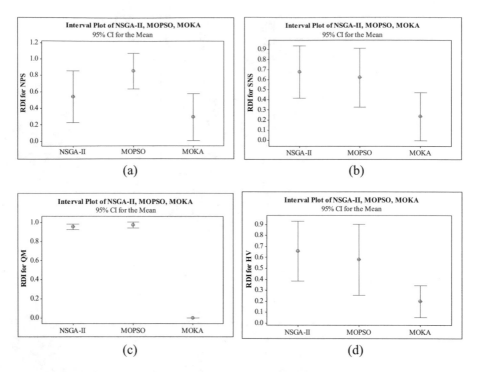

**Fig. 2.8** Statistical analyses based on the interval plot for assessment metrics, i.e., (**a**) NPS, (**b**) SNS, (**c**) QM and (**d**) HV

## 2.6   Conclusion and Future Works

Academically, operational HHC decision-making seeks to optimize the scheduling and routing activities of the caregivers to visit the patients, using a simplified objective function. However, in the developed countries where the management of aging population is of particular concern, a simplified approach to the HHC is failing to deliver satisfactorily all the real aspects of this practical optimization problem. The aim of this study is to develop a new HHCRSP that is practical and efficient. Practicality requires a multi-objective optimization process that accommodates the total cost as a structured logistics network by a set of pharmacies and laboratories as well as the greenhouse gas emissions of the logistics activities. Since the travel and services times are uncertain and play a significant role to adjust the HHC scheduling, a robust optimization model among the first studies in this research area was proposed. Efficiency requires an algorithm to better solve this complex problem that is robust and computationally manageable such as the proposed KA in this study.

The results of this study have demonstrated the viability of an HHC optimization with multiple objectives and complex constraints. A bi-objective robust optimization for the first time and a recently developed metaheuristic applied to HHCRSP were introduced. This study proposed three capable metaheuristics including NSGA-II, MOPSO, and the proposed MOKA to solve the proposed HHCRSP in an acceptable computational time. The MOKA performs especially well given the particular problem definition. The problem definition itself is still somewhat simplified relative to the full scope and dynamics of all potential HHCRSP in practice, but it does represent an order of magnitude more complexity than the vast majority of previous HHC definitions. Most notably, this study includes objective functions that accommodate the full scope of potential financial and sustainable aspects of the HHC management. The gas emissions and stochastic travel and service times were firstly contributed simultaneously to the problem by developing a robust optimization model. Last but not least, the efficiency of the MOKA in this context also lends great encouragement to the broader adoption and application of the approach in other problem domains of the equivalent compounding complexity.

The HHCRSP definition in this study revealed a new order of complexity for the study of patients' allocation as well as the routing and scheduling of the caregivers, but the possibility to introduce further sustainable factors is both necessary to reflect the full scope of HHC optimization in practice and a positive outcome of the current study. For example, more sustainability factors such as patients' risks and satisfaction associated with the HHC services would offer interesting additions. Further sensitivity analysis using other assessment metrics of Pareto-based algorithms could also be applied to the current model, and the introduction of more sensitivity analyses on the key parameters of the model as well as the effects of uncertain parameters on the proposed robust model would also extend the study in several dimensions. More broadly, and ultimately more significantly perhaps, the application and development of MOKA to other practical optimization problems such as humanitarian logistics or water distribution and wastewater collection offer exciting new directions for future researches.

# References

1. Decerle, J., Grunder, O., El Hassani, A. H., Barakat, O.: A memetic algorithm for a home health care routing and scheduling problem. Oper. Res. Health Care **16**, 59–71 (2018)
2. Tyan, M.: Understanding Taiwanese home healthcare nurse managers' empowerment and international learning experiences: community-based participatory research approach using a US home healthcare learning tour. University of Washington (2010)
3. Fathollahi-Fard, A.M., Hajiaghaei-Keshteli, M., Tavakkoli-Moghaddam, R.: A Lagrangian relaxation-based algorithm to solve a Home Health Care routing problem. Int. J. Eng. **31**(10), 1734–1740 (2018)
4. Decerle, J., Grunder, O., El Hassani, A. H., Barakat, O.: A hybrid memetic-ant colony optimization algorithm for the home health care problem with time window, synchronization and working time balancing. Swarm Evol. Comput. **34**, 283–295 (2019)
5. Fikar, C., Hirsch, P.: Home health care routing and scheduling: a review. Comput. Oper. Res. **77**, 86–95 (2017)

6. Ji, A.B., Qiao, Y., Liu, C.: Fuzzy DEA-based classifier and its applications in healthcare management. Health Care Manag. Sci. **22**, 560–568 (2019)
7. Fathollahi-Fard, A.M., Hajiaghaei-Keshteli, M., Mirjalili, S: A set of efficient heuristics for a home healthcare problem. Neural Comput. Appl. 1–21 (2019)
8. Fikar, C., Hirsch, P.: A matheuristic for routing real-world home service transport systems facilitating walking. J. Clean. Prod. **105**, 300–310 (2015)
9. Sahebjamnia, N., Fathollahi-Fard, A.M., Hajiaghaei-Keshteli, M.: Sustainable tire closed-loop supply chain network design: hybrid metaheuristic algorithms for large-scale networks. J. Clean. Prod. **196**, 273–296 (2018)
10. Hajiaghaei-Keshteli, M., Aminnayeri, M.: Keshtel Algorithm (KA); a new optimization algorithm inspired by Keshtels' feeding. In: Proceeding in IEEE Conference on Industrial Engineering and Management Systems, pp. 2249–2253 (2013).
11. Begur, S.V., Miller, D.M., Weaver, J.R.: An integrated spatial DSS for scheduling and routing home-health-care nurses. Interfaces **27**(4), 35–48 (1997)
12. Bertels, S., Fahle, T.: A hybrid setup for a hybrid scenario: combining heuristics for the home health care problem. Comput. Oper. Res. **33**(10), 2866–2890 (2006)
13. Akjiratikarl, C., Yenradee, P., Drake, P.R.: PSO-based algorithm for home care worker scheduling in the UK. Comput. Ind. Eng. **53**(4), 559–583 (2007)
14. Eveborn, P., Rönnqvist, M., Einarsdóttir, H., Eklund, M., Lidén, K., Almroth, M.: Operations research improves quality and efficiency in home care. Interfaces **39**(1), 18–34 (2009)
15. Trautsamwieser, A., Gronalt, M., Hirsch, P.: Securing home health care in times of natural disasters. OR Spectr. **33**(3), 787–813 (2011)
16. Rasmussen, M.S., Justesen, T., Dohn, A., Larsen, J.: The home care crew scheduling problem: preference-based visit clustering and temporal dependencies. Eur. J. Oper. Res. **219**(3), 598–610 (2012)
17. Liu, R., Xie, X., Augusto, V., Rodriguez, C.: Heuristic algorithms for a vehicle routing problem with simultaneous delivery and pickup and time windows in home health care. Eur. J. Oper. Res. 230(3), 475–486 (2013)
18. Liu, R., Xie, X., Garaix, T.: Hybridization of Tabu search with feasible and infeasible local searches for periodic home health care logistics. Omega **47**, 17–32 (2014)
19. Mankowska, D.S., Meisel, F., Bierwirth, C.: The home health care routing and scheduling problem with interdependent services. Health Care Manag. Sci. **17**(1), 15–30 (2014)
20. Braekers, K., Hartl, R.F., Parragh, S.N., Tricoire, F.: A bi-objective home care scheduling problem: analyzing the trade-off between costs and client inconvenience. Eur. J. Oper. Res. **248**(2), 428–443 (2016)
21. Shi, Y., Boudouh, T., Grunder, O.: A hybrid genetic algorithm for a home health care routing problem with time window and fuzzy demand. Exp. Syst. Appl. **72**, 160–176 (2017)
22. Lin, C.C., Hung, L.P., Liu, W.Y., Tsai, M.C.: Jointly rostering, routing, and re-rostering for home health care services: a harmony search approach with genetic, saturation, inheritance, and immigrant schemes. Comput. Ind. Eng. **115**, 151–166 (2018)
23. Liu, M., Yang, D., Su, Q., Xu, L.: Bi-objective approaches for home healthcare medical team planning and scheduling problem. Comput. Appl. Math. **37**(4), 4443–4474 (2018)
24. Cappanera, P., Scutellà, M.G., Nervi, F., Galli, L.: Demand uncertainty in robust Home Care optimization. Omega **15**, 623–635 (2018)
25. Fathollahi-Fard, A.M., Hajiaghaei-Keshteli, M., Tavakkoli-Moghaddam, R.: A bi-objective green home health care routing problem. J. Clean. Prod. **200**, 423–443 (2018)
26. Shi, Y., Boudouh, T., Grunder, O., Wang, D.: Modeling and solving simultaneous delivery and pick-up problem with stochastic travel and service times in home health care. Exp. Syst. Appl. **102**, 218–233 (2018)
27. Demirbilek, M., Branke, J., Strauss, A.: Dynamically accepting and scheduling patients for home healthcare. Health Care Manag. Sci. **22**(1), 140–155 (2019)
28. Shi, Y., Boudouh, T., Grunder, O.: A robust optimization for a home health care routing and scheduling problem with consideration of uncertain travel and service times. Transp. Res. E Logist. Transp. Rev. **128**, 52–95 (2019)

29. Bahadori-Chinibelagh, S., Fathollahi-Fard, A.M., Hajiaghaei-Keshteli, M.: Two constructive algorithms to address a multi-depot home healthcare routing problem. IETE J. Res. 1–7 (2019). https://doi.org/10.1080/03772063.2019.1642802
30. Grenouilleau, F., Legrain, A., Lahrichi, N., Rousseau, L.M.: A set partitioning heuristic for the home health care routing and scheduling problem. Eur. J. Oper. Res. 275(1), 295–303 (2019)
31. Ben-Tal, A., Nemirovski, A.: Robust solutions of uncertain linear programs. Oper. Res. Lett. 25(1), 1–13 (1999)
32. Fathollahi-Fard, A.M., Hajiaghaei-Keshteli, M., Tavakkoli-Moghaddam, R.: The social engineering optimizer (SEO). Eng. Appl. Artif. Intell. 72, 267–293 (2018)
33. Fathollahi-Fard, A.M., Govindan, K., Hajiaghaei-Keshteli, M., Ahmadi, A.: A green home health care supply chain: new modified simulated annealing algorithms. J. Clean. Prod. 240, 118–200 (2019)
34. Samanlioglu, F., Ferrell, W.G., Jr., Kurz, M.E.: A memetic random-key genetic algorithm for a symmetric multi-objective traveling salesman problem. Comput. Ind. Eng. 55(2), 439–449 (2008)
35. Hajiaghaei-Keshteli, M., Aminnayeri, M.J.A.S.C.: Solving the integrated scheduling of production and rail transportation problem by Keshtel algorithm. Appl. Soft Comput. 25, 184–203 (2014)
36. Fathollahi-Fard, A.M., Ranjbar-Bourani, M., Cheikhrouhou, N., Hajiaghaei-Keshteli, M.: Novel modifications of social engineering optimizer to solve a truck scheduling problem in a cross-docking system. Comput. Ind. Eng. 137, 106103 (2019)
37. Fathollahi-Fard, A.M., Hajiaghaei-Keshteli, M.: A stochastic multi-objective model for a closed-loop supply chain with environmental considerations. Appl. Soft Comput. 69, 232–249 (2018)
38. Golshahi-Roudbaneh, A., Hajiaghaei-Keshteli, M., Paydar, M.M.: Developing a lower bound and strong heuristics for a truck scheduling problem in a cross-docking center. Knowl.-Based Syst. 129, 17–38 (2017)
39. Safaeian, M., Fathollahi-Fard, A.M., Tian, G., Li, Z., Ke, H.: A multi-objective supplier selection and order allocation through incremental discount in a fuzzy environment. J. Intell. Fuzzy Syst. 37(1), 1435–1455 (2019). https://doi.org/10.3233/JIFS-182843
40. Fathollahi-Fard, A.M., Hajiaghaei-Keshteli, M., Mirjalili, S.: A set of efficient heuristics for a home healthcare problem. Neural Comput. Appl. 32 6185–6205 (2019)
41. Feng, Y., Zhang, Z., Tian, G., Fathollahi-Fard, A.M., Hao, N., Li, Z., Tan, J.: A novel hybrid fuzzy grey TOPSIS method: supplier evaluation of a collaborative manufacturing enterprise. Appl. Sci. 9(18), 3770 (2019)
42. Kim, S.H., Na, S.W.: Response surface method using vector projected sampling points. Struct. Safety 19(1), 3–19 (1997)
43. Fu, Y., Tian, G., Fathollahi-Fard, A.M., Ahmadi, A., Zhang, C.: Stochastic multi-objective modelling and optimization of an energy-conscious distributed permutation flow shop scheduling problem with the total tardiness constraint. J. Clean. Prod. 226, 515–525 (2019)
44. Haimes, Y.Y., Ladson, L.S., Wismer, D.A.: Bicriterion formulation of problems of integrated system identification and system optimization. IEEE Trans. Syst. Man Cybern. 3, 296 (1971)

# Chapter 3
# A Skewed General Variable Neighborhood Search Approach for Solving the Battery Swap Station Location-Routing Problem with Capacitated Electric Vehicles

**Mannoubia Affi, Houda Derbel, Bassem Jarboui, and Patrick Siarry**

**Abstract** The station location problem plays an important role for reducing the amount of energy consumption in several logistics companies while identifying the location of the stations and the number of the located stations. In addition, an electric vehicle represents a significant factor to minimize the costs and to reduce the pollution caused by transport operations. To achieve these goals, we consider a new variant of location-routing problem for an electric vehicle with a single depot, such as the location decision is undertaken about the battery swap stations with an elaboration of vehicle routing. Our solution method is based on a skewed version of the variable neighborhood search (SVNS) metaheuristic. The local search corresponds to the mixed variable neighborhood descent (Mixed-VND) where a set of four routing neighborhood structures are described in the sequential VND procedure and nested with relocated station move. The solution generated by the local search is compared to the current solution according to a particular distance function. The obtained results for all available instances of the problem show the performance of our approach compared to those existing in the literature.

M. Affi (✉) · H. Derbel
Faculty of Economic Sciences and Management of Sfax, Sfax, Tunisia

B. Jarboui
Department of Business, Higher Colleges of Technology, Abu Dhabi Women's College, Abu Dhabi, UAE

P. Siarry
Université Paris-Est Créteil Val-de-Marne, Laboratoire LiSSi, Créteil, France
e-mail: siarry@univ-paris12.fr

© Springer Nature Switzerland AG 2020                                          75
H. Derbel et al. (eds.), *Green Transportation and New Advances in Vehicle Routing Problems*, https://doi.org/10.1007/978-3-030-45312-1_3

## 3.1  Introduction

The reduction of the renewable energy consumption is the main objective of many companies. Many logistic problems are developed to deal with when and where to stop at the battery swap station for recharging the vehicle in order to decrease the cost of the total energy consumption. Different variants of location-routing problem (LRP) are presented to handle this objective. The LRP is a transportation logistic problem that involves two levels of decision: the location of depots (strategic level) and the development of the vehicle routing (tactical or operational level). In the literature, several studies have been carried out using exact and heuristic methods for solving the LRP problem. Among search studies in which the exact methods have been proposed, we can cite the work of [11] that studied an integer linear programming model for the LRP with only one depot and use the branch and cut method for solving it. Moreover, [2] developed a lower bound to a LRP with capacitated depots, uncapacitated vehicle, and only one route per depot. Among the proposed metaheuristics, [4] developed a hybrid method that combines genetic algorithm (GA) with an iterative local search (ILS) to solve the LRP problem with multiple capacitated depots and one uncapacitated vehicle per depot. Then, [10] proposed a general variable neighborhood search (GVNS) to solve the same problem and more recently in the work of [12], a skewed general variable neighborhood search (SGVNS) algorithm was proposed to solve the location-routing scheduling problem.

In this work, we deal with a variant of the LRP that has been considered by Yang and Sun [15]. It combines the location problem with green vehicle routing problem (G-VRP) to generate a new model that can be seen as a battery swap station location-routing problem with capacitated electric vehicles (BSS-EV-LRP). The G-VRP is a variant of the classical vehicle routing problem where a set of recharging stations should be integrated in the right time and placed for a given route in order to reduce the cost of the total energy consumption. Many G-VRP problems are proposed in order to reduce the cost of the total energy consumption. This type of problem was originally described by Erdoğan and Miller-Hooks [5]. Few studies have focused on the use of exact methods [5, 14] to solve the G-VRP. In contrast, several studies have focused on the development of heuristic and metaheuristic approaches [1, 3, 6, 8, 13] to solve the same variant of the G-VRP.

The difference between the standard LRP and BSS-EV-LRP is that the location decision is undertaken not about the depot but about the battery swap stations (BSS). Our work deals with the study of the problem combining the decision of the location of the set of BSSs and the elaboration of the vehicle routing. The battery swap station location problem is handled as a facility location problem where the discharged battery will be replaced by a new fully charged battery in the EV. The battery swap station offers a number of benefits over the refueling station including the reduced charge time: the time for switching batteries is much shorter than the refueling time. In addition, the BSS has an economic benefit: the electricity price to charge the empty batteries is lower than the refueling cost. Additionally, the use of BSS helps reduce air pollution: the electricity consumed by electric vehicles can be generated from tidal power, solar power, and wind power, which can reduce the consumption of nonrenewable energy (fuel, gasoline) and increase utilization of renewable energy (wind and solar resources). Hence, the greenhouse gas emission is reduced. The EV is also significantly quieter than conventional vehicles which leads to less noise pollution. To solve this variant, two approaches are proposed by Yang and Sun [15] to improve the quality of the solution: The first one is a hybrid Tabu search (TS) combined with modified Clarke and Wright savings (MCWS) method. Indeed, TS method is used to make the location decision and generate a new solution which is used as a starting point for the MCWS approach to search for a routing strategy. The second hybrid heuristic combines four-stage heuristic called SIGALNS where the BSSs location step and the vehicle routing step are affected separately.

The main contribution of this work is to propose an approach derived from variable neighborhood search algorithm for BSS-EV-LRP where a set of neighborhoods is used in the local search algorithm between the location and routing decisions. More precisely, we consider five neighborhood structures, four of which concern the routing decision and perform in two ways as follows: intra-route and inter-route, while the fifth neighborhood concerns the location decision (relocated station move). Our local search is presented by using the mixed-VND algorithm which has two main procedures: a relocated station move and a sequential VND. The solution generated by the local search is compared to the current solution according to a distance function. We disrupt the solution by deleting a customer node from its position and insert it at a random position. In fact, the experiment results demonstrated that our approach outperforms existing approach.

The remainder of this paper is structured as follows: The next section describes the problem. The third section provides the different steps of the proposed algorithm. Finally, we report and analyze our experimental results.

## 3.2   Problem Definition

The battery swap station location-routing problem with capacitated electric vehicles (BSS-EV-LRP) has been proposed by Yang and Sun [15]. It is defined by a graph $Z = (W, B)$, where $W$ is a total set of nodes and partitioned as $W = E \cup J \cup \{d\} \cup \{d'\}$. $E$ is a set of customers, $J$ is a set of BSSs without capacity constraints and nodes $d$, and $d'$ denote the original depot and the virtual depot, respectively. $B$ is the set of arcs $(e, f)$, $e, f \in E$. Each arc $(e, f)$ has a cost $d_{ef}$ which equals the distance required to travel this arc. The fleet consists of $m$ identical electric vehicles with a capacity $U_h$, $h = 1..m$ and a limited battery power level $L$. The service at each node $e$ consists to deliver a quantity of goods $q_e$ ($q_e = 0$ if node $e$ is a BSS). The electric vehicles deliver goods from depot to customers with given demands and can stop at BSSs to exchange their discharged battery for a new fully charged one. The objective function of the BSS-EV-LRP consists in minimizing the total cost including a fixed cost of building associated with the establishment of a station for each BSS candidate and a shipping cost of the unit associated with any tour traveled by the electric vehicles. The following constraints must be respected: (i) the depot must be the first and final visited node by each EV, (ii) each route must be started with a fully charged battery, (iii) the traveled distance between the depot and a visit to a BSS or two BSSs does not exceed the battery charge limit, (iv) the residual charge at EV at any node cannot violate the EV capacity, (v) each located BSS can be revisited more than once, and (vi) each customer should be served once by one EV.

## 3.3 Solution Approach

In this section, we present our solution approach based on the VNS metaheuristic. More precisely, we define a skewed general version of the VNS (SGVNS) approach by using a mixed variable neighborhood descent as local search where it combines nested VND with sequential VND. More details are provided in the following subsections.

### 3.3.1 Local Search

#### 3.3.1.1 Neighborhood Structures

Our method involves the application of various neighborhood structures in the local search algorithm monitoring the location and routing decisions. We have developed different types of moves within five neighborhood structures where one neighborhood concerns the location decision: $N_1$ relocated station move. While the other four neighborhoods were developed to undertake the routing decision into consideration. In fact, they are alternated between intra-route and inter-route moves. The four proposed routing neighborhoods are described as follows: $N_2$ : one-customer move, $N_3$ : two-customers move, $N_4$ : three-customers move, and finally $N_5$ : several-customers move.

- Location neighborhood
  - Relocated station move ($N_1$): This type of move searches for a neighbor of a given solution $x$. It basically replaces a located station with an unlocated or already a located one in a best position. In the case of battery with a high charge level, the station is removed. This operation is repeated for all the located stations. Whenever a relocated station move is realized, a new solution $x' \in N_1(x)$ is obtained. More precisely, as shown in Fig. 3.1, we select two stations $s_1$ and $s_2$ such that $s_1$ is located and $s_2$ is unlocated. Then,

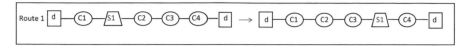

Fig. 3.1 Example of a move in neighborhood $N_1$

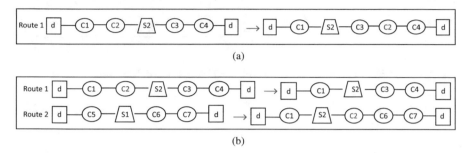

Fig. 3.2 Example of a move in neighborhood $N_2$. (a) Intra-route. (b) Inter-route

after removing station $s_1$ from its position, we insert station $s_2$ in the best position in terms of cost in order to make a good location decision.

• Routing neighborhoods

  – One-customer move ($N_2$): This type of move essentially relocates a selected customer from the current position into a new position in the same route (intra-route) or in another route (inter-route). In fact, within an intra-route move, $N_2$ consists of applying one-customer move by shifting a customer from position $i$ to position $j$ in the same route as shown in Fig. 3.2a. In the case of inter-route move, we shifted a selected customer in a different route, see Fig. 3.2b. After applying this move to all the customers and for all the possible positions to a given solution $x$, the best solution $x^{'} \in N_2$ is generated.

  – Two-customers move ($N_3$): This type of move consists of selecting two different customers and then exchanging their positions. The two selected customers can be chosen from the same route as shown in Fig. 3.3a or from two different routes as shown in Fig. 3.3b. After performing this change to all possible two customers from a solution $x$, a new neighboring solution $x^{'} \in N_3(x)$ is generated.

  – Three-customers move ($N_4$): A neighbor of a solution $x$ is obtained by selecting two adjacent customers and exchanging them with a customer belonging to another route as illustrated in Fig. 3.4b or in the same route as illustrated in Fig. 3.4a. This move searches for one-neighbor of a given solution $x$ to generate a new neighboring solution $x^{'}$.

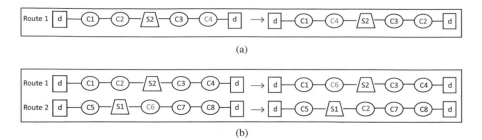

Fig. 3.3 Example of a move in neighborhood $N_3$. (a) Intra-route. (b) Inter-route

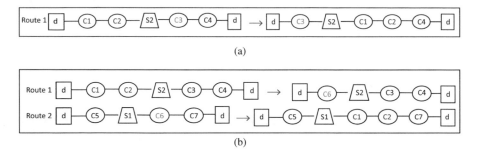

Fig. 3.4 Example of a move in neighborhood $N_4$. (a) Intra-route. (b) Inter-route

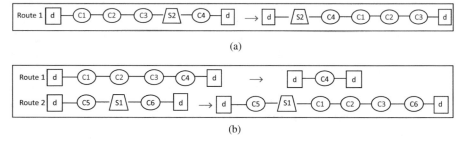

Fig. 3.5 Example of a move in neighborhood $N_5$. (a) Intra-route. (b) Inter-route

- Several-customers move ($N_5$): A neighbor of a solution $x$ is obtained by removing a sequence of consecutive customers from a current position and reinserting them in another position of the same route, see Fig. 3.5a, or of another route, see Fig. 3.5b. This move is repeated for all the customers and for all the available positions.

### 3.3.1.2   Seq-VND

The sequential VND (Seq-VND) is defined for the first time by Hansen et al. [7], when the change of neighborhoods is sequential. The algorithm takes two input components: a current solution $x$ and the maximum number of used neighborhood structures $i_{max}$. Seq-VND iteratively explores each of the considered neighborhoods according to a sequential order by returning to the first neighborhood at each time a better solution can be found. The method stops if it is no longer possible to generate a better solution over a set of $i_{max}$ neighborhood structures. To apply our algorithm, the input parameter $i_{max}$ corresponds to the number of the routing neighborhood structures described in Sect. 3.3.1.1. Then, at each iteration, the current solution is improved by exploring our proposed routing neighborhood $N_i$ where $N_i \in \{N_2, N_3, N_4, N_5\}$. Different steps of the Seq-VND are described in Algorithm 1.

### 3.3.1.3   Mixed-VND

Our local search corresponds to the mixed variable neighborhood descent algorithm (Mixed-VND) which was firstly introduced by [9] to solve the p-hub median problem. We apply the neighborhood $N_1$ nested with the four neighborhoods $N_2, N_3, N_4, N_5$ in a nested way as follows: let $J_l$ be the set of located BSSs in an initial solution $x$ and $J_u = J \setminus J_l$ be the set of unlocated ones in $x$. Then, our algorithm is presented by applying a relocated station move $N_1$ for a randomly selected BSS in $J_l$ and then a Seq-VND. For more details, we firstly changed the incumbent solution by applying relocated station move in $N_1(x)$ to get a new feasible neighbor $x^{'}$, which is used as an input to the Seq-VND. In case of no improvement of the solution $x$, the algorithm stops.

Algorithm 2 describes more clearly the process of the mixed-VND. Parameter $i_{max}$ is the number of used neighborhood structures in Seq-VND.

---

**Algorithm 1:** Seq-VND $(x)$

---

**Input:**
  $R = \{N_2, N_3, N_4, N_5\}$ ;
  $i_{max} = |R|$ ;
  $i = 1$ ;

1 **repeat**
2     **if** *(i = 1)* **then**
3        $x' =$ the first improvement on using neighborhood $N_2$ ;
4     **else if** *(i = 2)* **then**
5        $x' =$ the first improvement on using neighborhood $N_3$ ;
6     **else if** *(i = 3)* **then**
7        $x' =$ the first improvement on using neighborhood $N_4$ ;
8     **else if** *(i = 4)* **then**
9        $x' =$ the first improvement on using neighborhood $N_5$ ;
10     **if** $(f(x') < f(x))$ **then**
11        $x := x'$ ;
12        $i := 1$;
13     **else**
14        $i + +$ ;
15 **until** $(i = i_{max})$;
16 **return** $x$;

---

 

---

**Algorithm 2:** Mixed-VND $(x)$

---

**Input:**
  Let $J_l$ be the set of BSSs located in an initial solution $x$;
  Let $J_u$ be the set of BSSs unlocated in an initial solution $x$;

1 **repeat**
2     Select a random neighbor $x' \in N_1(x)$ ;
3     $x'' := $ Seq-VND$(x')$ ;
4     **if** $(f(x'') < f(x))$ **then**
5        $x := x''$;
6     Update $J_l$ in the current solution $x$;
7     Update $J_u$ in the current solution $x$;
8 **until** *no possible improvement*;
9 **return** $x$;

---

### 3.3.2  Skewed General Variable Neighborhood Search Algorithm (SGVNS)

The SGVNS is summarized in Algorithm 3 where it needs the following main elements: a random generated solution $x$, a set of five neighborhood structures, a stopping criterion, mixed-VND algorithm, and a shaking phase. Parameter $t_{max}$ is the maximum time allocated to the algorithm and $i_{max}$ is the maximum number of consecutive disturbances to apply to the current solution. After generating the solution $x$, the shaking phase begins to give a new neighboring solution $\bar{x}$. In the shaking phase, the incumbent solution is perturbed by removing a customer node from its position and inserting it randomly. Then, the mixed-VND algorithm is used to look for a good exploration of our proposed neighborhoods in the solution $\bar{x}$ to give a new solution $x^+$ which is compared to the best solution $x^*$. Solution $x^+$ is accepted if its objective value is better. Solution $x^+$ is also compared to the current solution according to a distance function (see the next subsection "Distance Function" for more details). We accept the solution $x^+$ as the current solution if it is better. Then, the algorithm goes back to the shaking phase. SGVNS ends when a time limit $t_{max}$ is reached.

---

**Algorithm 3:** SGVNS $(i_{max}; t_{max})$

---

   **Input:**
      $N = \{N_{j \in \{1,...,5\}}\}$ : a set of neighborhood structures
      $x$ : a randomly generated solution;
      $x^* = x$ ;

1  **repeat**
2     $i = 1$ ;
3     **repeat**
4         $\bar{x} :=$ Shaking$(x)$ ;
5         $x^+ := $ Mixed-VND $(\bar{x}, N_{j \in \{1,...,5\}})$;
6         **if** $(f(x^+) < f(x^*))$ *and* $x^+$ *is feasible* **then**
7             $x^* := x^+$;
8         **if** $f(x^+) < (1 + \mu\rho(x, x^+))f(x)$ **then**
9             $x := x^+$ ;
10            $i := 1$ ;
11         **else**
12            $i + +$;
13     **until** $(i > i_{max})$;
14 **until** $(t > t_{max})$;
15 **return** $x$;

---

The generated solution from the local search is compared to the current solution according to a distance function. This distance is used to allow to accept a worse solution. In this case, the distance function value $\rho(x, x^+)$ is computed by taking into account the subset of located stations in the current solutions $x$ and the generated one $x^+$. In fact, we define the two following parameters:

$$\delta_j = \begin{cases} 1, & \text{if station } j \text{ is located at } x \\ 0, & \text{otherwise} \end{cases} \tag{3.1}$$

$$\delta_j^+ = \begin{cases} 1, & \text{if station } j \text{ is located at } x^+ \\ 0, & \text{otherwise} \end{cases} \tag{3.2}$$

Then, the distance between both solutions $x^+$ and $x$ is computed as follows:

$$\rho(x, x^+) = \frac{\sum_{j=1}^{J} |\delta_j - \delta_j^+|}{m} \tag{3.3}$$

Hence, the generated solution $x^+$ is accepted to be the current solution if:

$$f(x^+) < (1 + \mu\rho(x, x^+))f(x) \tag{3.4}$$

where $\mu$ is a constant parameter and $m$ is the number of vehicles.

## 3.4  Experimental Results

Our SGVNS algorithm is implemented in C ++ language and tested with an Intel (R) Core (TM) i5-4460 CPU, 3.20GHz. To evaluate the quality of the proposed SGVNS method, we compare our results to those generated by SIGALNS algorithm proposed by Yang and Sun [15]. Both compared approaches are run 10 times on each instance of the literature. Our approach requires the adjustment of some parameters. The three parameters of the method are $i_{max}$, $t_{max}$, and $\mu$, where $i_{max}$ is the maximum number of iteration in the shaking phase, $t_{max}$ is the maximum time allocated to the SGVNS algorithm, and $\mu$ is a constant parameter. The values used for these parameters are $i_{max} = 10$, $t_{max} = $ n*10 seconds, where n is the number of customers and $\mu$=0.1.

Tables 3.1 and 3.2 present the results given by SIGALNS algorithm and our proposed method SGVNS for the BSS-EV-LRP problem. The first column, instance, reports the name of the instance, while the column BKS gives the best-known solution. For SIGALNS and SGVNS algorithm, $k^*$ is the best found solution obtained after 10 runs, $h$ is the number of used vehicles in the best found solution,

$nb$ is the number of located stations in the best found solution, $k_a$ is the average solution obtained after 10 runs, $g(\%)$ is the deviation between the BKS and the best solution, $g_a(\%)$ is the deviation between the BKS and the average solution, and $t$ is the running time in seconds. In the last two lines, we present the averages of the running time, the average percentage deviation, and the total number of the located stations. Namely, the best solutions are marked in bold. The best solutions, the best number of used vehicles, the best number of located stations obtained after 10 runs are marked in bold.

The minimum percentage deviation $g(\%)$ of the ten runs is given by

$$\frac{(k^* - BKS)}{BKS} * 100$$

and the average percentage deviation $g_a(\%)$ relative to the BKS of the ten runs is measured by

$$\frac{(k_a - BKS)}{BKS} * 100$$

By examining Table 3.1, we note that our SGVNS reaches the best solution value for all the very small instances and the running time of our method is on average smaller than 2 s which is very small.

After analyzing Table 3.2, the results show that the best solutions have been produced by our proposed method when compared with SIGALNS heuristic. The solution values of our algorithm always perform better for all instances. Moreover, we can improve the number of located BSSs for 14 instances out of 24. Then, we can improve the total number of the located stations where it is 88 for all instances of the SIGALNS approach and 30 for all instances of the SGVNS algorithm. The SGVNS can also improve the number of used vehicles for 4 instances out of 24. It also appears that our proposed method is always better than the SIGALNS with respect to the solution quality and by comparing the total average deviation obtained by the SGVNS algorithm ($-6.91\%$ and $-3.41\%$) with the SIGALNS ($0.62\%$). Moreover, the average running time is 62.322 s for SIGALNS and 53.919 s for SGVNS. Finally,

**Table 3.1** Comparison between the proposed approach and SIGALNS on the very small instances

| Instance | BKS | SIGALNS | | | SGVNS | | |
|---|---|---|---|---|---|---|---|
| | | $k^*$ | $g(\%)$ | $t(s)$ | $k^*$ | $g(\%)$ | $t(s)$ |
| P-n6-k2 | 426.86 | 426.86 | 0.00 | 1.73 | 426.86 | 0.00 | 0.07 |
| P-n7-k3 | 428.6 | 428.6 | 0.00 | 1.93 | 428.6 | 0.00 | 0.38 |
| P-n8-k3 | 597.16 | 597.16 | 0.00 | 2 | 597.16 | 0.00 | 1.08 |
| RY-n12-k2 | 52792.61 | 52792.61 | 0.00 | 1.91 | 52792.61 | 0.00 | 3.52 |
| RY-n15-k3 | 52985.62 | 52985.62 | 0.00 | 1.86 | 52985.62 | 0.00 | 1.15 |
| RY-n20-k4 | 63919.2 | 63919.2 | 0.00 | 2.22 | 63919.2 | 0.00 | 1.23 |
| Tot-Avg | | | 0.00 | 1.94 | | 0.00 | 1.22 |

**Table 3.2** Comparison between the proposed approach and SIGALNS on the small, medium, and large instances

| Instance | BKS | SIGALNS | | | | | SGVNS | | | | | | |
|---|---|---|---|---|---|---|---|---|---|---|---|---|---|
| | | $k^*$ | $h$ | $nb$ | $g(\%)$ | $t(s)$ | $k^*$ | $h$ | $nb$ | $k_a$ | $g(\%)$ | $g_a(\%)$ | $t(s)$ |
| P-n16-k8 | **1281.95** | **1281.95** | 8 | 1 | 0.00 | 2.58 | **1281.95** | 8 | 1 | 1321.08 | 0.00 | 3.05 | 0.49 |
| P-n19-k2 | 471.39 | 471.39 | 2 | 1 | 0.00 | 2.78 | **466.18** | 2 | 1 | 470.77 | −1.11 | −0.13 | 1.88 |
| P-n21-k2 | 478.64 | 478.64 | 2 | 1 | 0.00 | 3.13 | **472.74** | 2 | 1 | 472.74 | −1.23 | −1.23 | 1.93 |
| P-n23-k8 | 1360.51 | 1360.51 | 8 | 1 | 0.00 | 3.14 | **1352.77** | 8 | 1 | 1377.84 | −0.57 | 1.27 | 6.72 |
| P-n40-k5 | 893.23 | 893.23 | 5 | 1 | 0.00 | 6.18 | **889.00** | 5 | 1 | 912.26 | −0.47 | 2.13 | 2.82 |
| P-n45-k5 | 939.63 | 939.63 | 5 | 2 | 0.00 | 7.69 | **909.74** | 5 | **1** | 951.06 | −3.18 | 1.22 | 2.95 |
| P-n50-k7 | 1196.48 | 1196.48 | 7 | 2 | 0.00 | 8.52 | **1194.48** | 7 | 2 | 1232.32 | −0.17 | 3.00 | 2.72 |
| P-n55-k8 | 1247.10 | 1247.10 | 7 | 2 | 0.00 | 20.13 | **1211.13** | 7 | 2 | 1267.43 | −2.88 | 1.63 | 13.52 |
| P-n60-k10 | 1684.24 | 1684.24 | 10 | 3 | 0.00 | 24.50 | **1647.16** | 10 | 3 | 1693.95 | −2.20 | 0.58 | 22.33 |
| P-n70-k10 | 1738.98 | 1738.98 | 10 | 3 | 0.00 | 35.93 | **1736.90** | 10 | 3 | 1761.24 | −0.12 | 1.28 | 33.82 |
| Average | | | | | 0 | 11.46 | | | | | **−1.19** | **1.28** | **8.73** |
| tai_75a | 1924.32 | 1924.32 | 10 | 4 | 0.00 | 53.69 | **1705.54** | 10 | **0** | 1835.09 | −11.37 | −4.64 | 14.48 |
| tai_75b | 1607.22 | 1607.22 | 10 | 3 | 0.00 | 76.73 | **1482.36** | 10 | **0** | 1513.50 | −7.77 | −5.83 | 14.45 |
| tai_75c | 1602.15 | 1602.15 | 9 | 4 | 0.00 | 75.11 | **1431.05** | 9 | **0** | 1516.47 | −10.68 | −5.35 | 11.33 |
| tai_75d | 1643.63 | 1643.63 | 9 | 3 | 0.00 | 51.06 | **1404.86** | 9 | **0** | 1500.51 | −14.53 | −8.71 | 25.96 |
| tai_100a | 2467.90 | 2467.90 | 12 | 4 | 0.00 | 118.8 | **2265.56** | **11** | **1** | 2332.18 | −8.20 | −5.50 | 51.34 |
| tai_100b | 2393.34 | 2393.34 | 12 | 5 | 0.00 | 134.42 | **1944.86** | **11** | **0** | 2023.11 | −18.74 | −15.47 | 31.65 |
| tai_100c | 1683.69 | 1783.45 | 11 | 4 | 5.93 | 123.29 | **1594.00** | 11 | **1** | 1649.64 | −5.33 | −2.02 | 82.23 |
| tai_100d | 1918.81 | 1926.96 | 12 | 4 | 0.42 | 188.95 | **1733.79** | 12 | **1** | 1767.14 | −9.64 | −7.90 | 88.50 |
| tai_150a | 3620.34 | 3620.34 | 15 | 4 | 0.00 | 329.67 | **3377.13** | 15 | **1** | 3510.54 | −6.72 | −3.03 | 80.80 |
| tai_150b | 3354.00 | 3354.00 | 14 | 7 | 0.00 | 367.24 | **2863.81** | 14 | **0** | 3109.84 | −14.61 | −7.28 | 319.25 |
| tai_150c | 2879.32 | 2879.32 | 15 | 5 | 0.00 | 337.44 | **2475.28** | 15 | **0** | 2608.47 | −14.03 | −9.41 | 230.61 |
| tai_150d | 3121.36 | 3121.36 | 15 | 5 | 0.00 | 492.06 | **2839.23** | **14** | **0** | 2991.43 | −9.04 | −4.16 | 214.78 |

| | | | | | | | | | | | | | |
|---|---|---|---|---|---|---|---|---|---|---|---|---|---|
| Average | | | | | 0.53 | 195.71 | | | | | **-10.89** | -6.61 | **97.11** |
| GW_09 | 787.52 | 790.99 | 14 | 5 | 0.44 | 1798.23 | **739.68** | **14** | 8 | 778.40 | -6.08 | -1.16 | 994.65 |
| GW_16 | 2182.47 | 2359.08 | 38 | 14 | 8.09 | 10695.97 | **1809.93** | **37** | 2 | 1874.57 | -17.07 | -14.11 | 10693.38 |
| Average | | | | | 4.27 | 6247.10 | | | | | **-11.57** | -7.63 | **5844.02** |
| tot-Avg | | | | | 0.62 | 623.22 | | | | | **-6.91** | -3.41 | **539.19** |
| Total | | | | 88 | | | | | **30** | | | | |

based on these results we can conclude that SGVNS algorithm outperforms the SIGALNS algorithms on five important criteria: the best solution value, the number of used vehicles, the number of located stations, the average deviation, and the time needed to converge towards the best solution.

## 3.5 Conclusion

In this work, we study the electric vehicles battery swap stations location and routing problem (BSS-EV-LRP) by tackling the two decisions: location and routing in a simultaneous way. We have been interested in proposing a new solution approach based on VNS to solve a variant of the LRP problem that seeks to find the optimal location decision of a set of available stations and the routing plan of electric vehicles simultaneously while considering the battery power level limitation. We propose different location and routing neighborhoods in a skewed version of VNS metaheuristic to solve the problem. We developed variable neighborhood structures in both local search and shaking phases that provide the interrelated nature about location and routing decisions. We compared our results with those provided by a heuristic called SIGALNS which used a modified sweep heuristic for initialization, iterative greedy method for location subproblem, adaptive large neighborhood search (ALNS) for routing subproblem, and finally an improvement procedure. Experimental results show the performance of our approach for all the available instances.

## References

1. Affi, M., Derbel, H., Jarboui, B.: Variable neighborhood search algorithm for the green vehicle routing problem. Int. J. Ind. Eng. Comput. 9(2), 195–204 (2018)
2. Albareda-Sambola, M., Dıaz, J.A., Fernández, E.: A compact model and tight bounds for a combined location-routing problem. Comput. Oper. Res. 32(3), 407–428 (2005)
3. Bruglieri, M., Pezzella, F., Pisacane, O., Suraci, S.: A variable neighborhood search branching for the electric vehicle routing problem with time windows. Electron. Notes Discrete Math. 47, 221–228 (2015)
4. Derbel, H., Jarboui, B., Hanafi, S., Chabchoub, H.: Genetic algorithm with iterated local search for solving a location-routing problem. Exp. Syst. Appl. 39(3), 2865–2871 (2012)
5. Erdoğan, S., Miller-Hooks, E.: A green vehicle routing problem. Transp. Res. E Logist. Transp. Rev. 48(1), 100–114 (2012)
6. Felipe, Á., Ortuño, M.T., Righini, G., Tirado, G.: A heuristic approach for the green vehicle routing problem with multiple technologies and partial recharges. Transp. Res. E Logist. Transp. Rev. 71, 111–128 (2014)
7. Hansen, P., Mladenović, N., Pérez, J.A.: Variable neighbourhood search: methods and applications. 4OR 6(4), 319–360 (2008)
8. Hiermann, G., Puchinger, J., Ropke, S., Hartl, R.F.: The electric fleet size and mix vehicle routing problem with time windows and recharging stations. Eur. J. Oper. Res. 252(3), 995–1018 (2016)

9. Ilić, A., Urošević, D., Brimberg, J., Mladenović, N.: A general variable neighborhood search for solving the uncapacitated single allocation p-hub median problem. Eur. J. Oper. Res. **206**(2), 289–300 (2010)
10. Jarboui, B., Derbel, H., Hanafi, S., Mladenović, N.: Variable neighborhood search for location routing. Comput. Oper. Res. **40**(1), 47–57 (2013)
11. Laporte, G., Nobert, Y.: An exact algorithm for minimizing routing and operating costs in depot location. Eur. J. Oper. Res. **6**(2), 224–226 (1981)
12. Macedo, R., Alves, C., Hanafi, S., Jarboui, B., Mladenović, N., Ramos, B., de Carvalho, J.M.V.: Skewed general variable neighborhood search for the location routing scheduling problem. Comput. Oper. Res. **61**, 143–152 (2015)
13. Schneider, M., Stenger, A., Goeke, D.: The electric vehicle-routing problem with time windows and recharging stations. Transp. Sci. **48**(4), 500–520 (2014)
14. Taha, M., Fors, M.N., Shoukry, A.A.: An exact solution for a class of green vehicle routing problem. In: International Conference on Industrial Engineering and Operations Management, pp. 7–9 (2014)
15. Yang, J., Sun, H.: Battery swap station location-routing problem with capacitated electric vehicles. Comput. Oper. Res. **55**, 217–232 (2015)

# Chapter 4
# The Cumulative Capacitated Vehicle Routing Problem Including Priority Indexes

**Karina Corona-Gutiérrez, Maria-Luisa Cruz, Samuel Nucamendi-Guillén, and Elias Olivares-Benitez**

**Abstract** This chapter studies the Cumulative Capacitated Vehicle Routing Problem including Priority Indexes, a variant of the classical Capacitated Vehicle Routing Problem, which serves the customers according to a certain level of preference. This problem can be effectively implemented in commercial and public environments where green concerns are incorporated, (like the reduction of $CO_2$ emission and energy consumption), and waste collection systems. For this problem, we aim to minimize two objectives: the total latency and the total tardiness of the system. A Mixed-Integer formulation is developed and solved using the AUGMECON approach to obtain true efficient Pareto fronts. However, as expected, the use of commercial software was able to solve only small instances, up to 15 customers. Therefore two metaheuristics were developed to solve the problem, one based on the Non-dominated Sorting Genetic Algorithm (NSGA) and the other based on Particle Swarm Optimization (PSO). These algorithms were used to solve the small instances where True Efficient Fronts were available. Both algorithms provided good solutions, although the NSGA algorithm obtained a better and denser Pareto front. Later, both algorithms were used to solve larger instances with 20–100 customers. The results were mixed in terms of quality, but the PSO algorithm performed faster. The instances solved were modified from benchmarks available in the literature. However, we are convinced that the model and algorithms proposed can be useful to solve a wide variety of situations where economic, environmental, and social concerns are involved.

K. Corona-Gutiérrez · M.-L. Cruz · S. Nucamendi-Guillén (✉) · E. Olivares-Benitez
Universidad Panamericana, Facultad de Ingeniería, Zapopan, Jalisco, Mexico
e-mail: kcorona@up.edu.mx; mlcruz@up.edu.mx; snucamendi@up.edu.mx; eolivaresb@up.edu.mx

© Springer Nature Switzerland AG 2020
H. Derbel et al. (eds.), *Green Transportation and New Advances in Vehicle Routing Problems*, https://doi.org/10.1007/978-3-030-45312-1_4

## 4.1 Introduction

Nowadays, the worldwide trend in terms of processes and the technologies associated focus on continuous research and improvement of efficiency in order to achieve and meet customer expectations. Regarding this, one of the most critical points to consider is the ability to quickly respond to customers' needs, whether in a commercial environment, service, or emergency. To get the necessary supplies to their destinations and to do it in the shortest possible time, will result in considerable benefits for those who provide services. This quick response implies to focus efforts and actions on what the customer requires and not just what the company needs. The ability to quickly respond to what customers need relates directly to the level of service quality, which by the customer's point of view represents the main driver for differentiation.

Customers' needs and expectations drive companies' strategic plans and activities. As stated before, one of their main concerns implies having immediate and quick response to their necessities. As a consequence of the imminent escalating deterioration of the environment, another primary concern of the customers relates to finding environment-friendly products and services. Thus, companies must take this into account in their decision-making process. While the industries' efforts commonly focus only on obtaining higher profits for companies, changing this approach and prioritizing customers' needs will inevitably result in increased satisfaction, which improves sales and then profits.

Companies have started to realize that the implementation of eco-friendly or green practices can positively impact on their activities from an economical environmental, and social point of view. They are also required by governments and legislations to monitor the impacts they have on the environment. This situation has led companies to realize that green practices are not just something they have to do but a vital philosophy to adapt and live within their daily activities.

Suposse we define a supply chain as a network between a company and its suppliers to produce and distribute a product to the final customer. In that case, a series of different processes, people, entities, and resources are required to achieve it. In the same way, we can define a green supply chain as an "integrating environmental thinking into supply-chain management, including product design, material sourcing, and selection, manufacturing processes, delivery of the final product to the consumers as well as end-life management of the product after its useful life" [1].

In order to get products to its final consumer, within the value chain, logistics related activities are necessary, and transportation is the central part of them. Engin et al. [2] state that transportation represents one-third of the logistics costs, becoming the most visible aspect of the supply chain. It is natural that logistics and transportation activities take the primary attention in the management of companies.

Due to the importance that transportation should represent on the strategic plan of any business and the needs of their customers, in order to fulfill those requirements, companies demand dynamic and effective solutions whose costs are efficient and can be applied in real-time. This subject is then suitable to be studied through operations research, by adapting the real problem to known approaches or by proposing new ones and getting appropiate and feasible tools to solve the problems that arise.

In particular, we can approach the described situation from the perspective of Vehicle Routing Problems (VRP), which traditionally seek to determine the optimal set of routes for a fleet of vehicles to traverse in order to deliver goods to a group of customers and focus on achieving the objective of a minimum cost.

However, when consider the importance that environmental issues should have in the decisions, besides the traditional objective of the VRP, the adverse effects of routing on the environment should be minimized, along with the economic costs of the routes. The main environmental concerns nowadays are related to global warming and the emission of greenhouse gases (GHGs) being the last one the most critical side effect of vehicle transportation, particularly the emission of carbon dioxide ($CO_2$).

So, in order to reduce the effects that transportation activities have on the environment, it is necessary to measure the $CO_2$ emissions and for that it will be better to know about the fuel consumption of the vehicles that carry out the transport and look for a way to minimize it. This problem has been addressed as a variant of the VRP under the name of Green Vehicle Routing Problems (GVRP). The GVRP focuses on minimizing the adverse effects on the environment, finding routes to satisfy the demand with a minimum economic cost. Its objective extends, taking into consideration an environmental aspect, either energy consumption, pollution, fuel consumption, or $CO_2$ emissions being the last two the main ones.

We also mentioned that one of the main concerns of customers focuses on their need for a quick response. So, companies have to worry about not putting aside the importance of customer satisfaction by minimizing the customer waiting time. This approach is known as the Cumulative Capacitated Vehicle Routing Problem (CCVRP) when planning a transportation strategy.

Several authors have previously addressed the CCVRP. The problem is a generalization of the $k$-Traveling Repairmen Problem ($k$-TRP), first introduced by Ngueveu et al. [3] to address problems in which the need for fast and fair services, customers related measures, becomes crucial. Since then, different scenarios derived from this problem have been analyzed: where a single vehicle can travel multiple trips [4, 5], considering stochastic demand and split/unsplit deliveries [6] or when multiple depots are available [7]. Previously, Kara et al. [8] studied a particular version of the CCVRP, named as CumVRP, in which the objective consists of minimizing the sum product of the arrival times and the demand of the node. Additionally, the CCVRP has been mainly studied from the mono-objective function perspective. For this, several contributions involving mathematical models [9, 10], exact algorithms [9, 11], and heuristic and metaheuristic approaches [3, 5, 12–14] have been developed.

The proposed models feature solutions that set up routes that consider only the minimum latency (total waiting time) without any other factor involved in the solution.

Other elements may have significance in the decision-making, and whose effect on the analysis of the problem, tools development, and solution procedures need to be considered. One such aspect is the priority to serve each customer. Each company should decide which aspects are essential to take into account in order to assign a priority level to its customers, either under the economic, social, environmental, or strategic points of view.

So, in this work, the Cumulative Capacitated Vehicle Routing Problem will be analyzed. However, unlike existing literature, the contribution of this research will be the introduction to the problem of a priority factor on customers. This priority will have an impact on the construction of the routes in order to meet customer demands. We must remember that the CCVRP aims to minimize the total latency of the system, so although it considers the priority, it cannot neglect the fact that a minimum waiting time has to be achieved. This problem can have practical applications in contexts where customers are willing to pay an extra fee to be served in minor time respect of the customers that did not pay for it. Also, there are different levels of preference to be selected, each with a separate fee. Another application can be in the context where perishable products (meat and vegetables) are delivered by the same vehicle to different customers.

The proposed model provides a bi-objective approach that takes care of the total waiting time of the system while ensuring that the customers' priority is respected (as far as possible). We measure the latter objective through a tardiness index, which represents the delay in the arrival time to a customer when its priority over the other customers is not respected. As in the total waiting time, this delay time is intended to be minimized.

The proposed CCVRP approach help to prioritize how to serve the customers in order to reduce their waiting time and increase their satisfaction, but through its construction and results also have an impact on environmental issues.

We stated before that each company will have different criteria when it comes to establishing a priority level for its customers, depending on the factors they consider essential. These factors might relate to the environment and impact either energy or fuel consumption or the amount of emissions generated. One of the elements to consider in this area is the distance traveled to the client since it will directly impact on the fuel required to do the transport. Another element is the demand to be satisfied as the number of products will represent more weight and hence more energy consumption. The type and size of vehicles will also impact fuel consumption and the overall weight moved on the tour. The roads available affect the emissions generated since lousy road conditions and driving in congested areas increase fuel consumption because the vehicles are riding under the optimal speed. Another element is the type of product. If we talk about perishable goods, the energy required to move and maintain them in optimal conditions will increase.

On the other hand, the system's total latency will also represent an environmental impact since the more extended time traveled, the bigger the energy and fuel consumed. Also, a delay in arrival to the client will increase the time the product spends in a vehicle, hence more energy needed in the move and storage.

The remainder of this paper is organized as follows: Sect. 4.2 presents the literature review regarding green VRPs. Section 4.3 formally describes the problem under study, whereas Sects. 4.4 and 4.5 describe the mathematical formulation and metaheuristic algorithms developed. Section 4.6 reports the computational experience and discusses the most relevant results. Finally, Sect. 4.7 summarizes the relevant insights and proposes future research lines.

## 4.2   Literature Review

The Green Vehicle Routing Problem is an extension of the traditional Vehicle Routing Problem covering broad objectives centered around environmental issues. The VRP focuses on the impact that routing costs have on logistics and, in particular, in the transportation activities within the supply chain. Due to the importance that sustainable practices have nowadays, there is a need to develop distribution strategies that help to reduce the negative impact that transportation activities have on the environment.

Lin et al. [15] define three significant categories of GVRP depending on the objective, and operational constraints considered in the problem: Green-VRP (GVRP), Pollution-Routing Problem (PRP), and VRP in Reverse Logistics (VRPRL). The research on GVRP deals with the optimization (minimization) of energy consumption of transportation in order to reduce greenhouse gas emissions in a significant way. The PRP aims to choose a vehicle dispatching scheme with less pollution, particularly reducing carbon emissions. The VRPRL focuses on the distribution aspects of reverse logistics.

Like in the traditional VRP and its known variants, two main approaches are used to solve the GVRP: exact solution methods and heuristic and metaheuristic algorithms. In their research, Karagul et al. [16] point out that optimal solutions are possible to obtain in small-scale problems using exact solution methods. However, large-scale problems are difficult and time-consuming to solve to optimality. Then, when it comes to real-life optimization where problems are complex and deal with a significant amount of data, sometimes it is enough to find approximate solutions through heuristic and metaheuristic methods.

Kara et al. [17] define the Energy-Minimizing Vehicle Routing Problem (EMVRP) as a variation of the CVRP with a new load-based cost objective. This cost function is a product of the vehicle's total load, and the length of the arcs traveled. Integer linear programming formulations with $O(n^2)$ binary variables and $O(n^2)$ constraints are used to solve both collection and delivery cases of the problem. The proposed models are tested on two CVRP instances taken from the literature. Results obtained show that there is a difference between energy and distance minimizing solutions. In general, energy usage increases as total distance decreases.

A computer-based vehicle routing model was developed by Palmer [18]. The model calculates and minimizes the overall amount of $CO_2$ generated on road journeys along with the total time and distance through the use of vehicle routing heuristics to construct the routes. Considering time windows and different traffic volumes, it explores how speed reduce emissions. The results show there is a potential to reduce the $CO_2$ emissions by about 5%.

In their research, Sbihi and Eglese [19] deal with vehicle routing models and issues relating to Green Logistics. They established that the use of time-dependent vehicle routing models (TDVRP) could provide solutions that produce less pollution by directing vehicles to roads where they can travel faster, which means away from congestion zones. However, this could represent traveling longer distances. Another benefit of this model is that it allows time window constraints to be satisfied more reliably.

Kuo [20] proposed a model for calculating total fuel consumption, total transportation time, and total distance traveled for the time-dependent vehicle routing problem (TDVRP) where speed and travel times are assumed to depend on the time of travel. The model also considers the loading weight. A simulated annealing (SA) algorithm is used to find the lowest total fuel consumption routes. The results show a trade-off between fuel consumption, transportation times, and transportation distances: the lower the fuel consumption, the longer the travel times and distance.

Maden et al. [21] addressed the vehicle routing and scheduling problem to minimize the total travel time. They considered that the time required for a vehicle to travel through the network varies depending on the travel time because of the congestions presented during rush hours. They proposed a Tabu Search based heuristic algorithm to solve the problem described in a case study. The results of the study show that using the proposed approach can lead to reductions in $CO_2$ emissions of about 7% when compared to the case with constant speeds. However, the model's objective remains to minimize the total travel time and not the reduction of emissions.

The Emissions Vehicle Routing Problem (EVRP) is presented by Figliozzi [22] as a variation of the Vehicle Routing Problem with Time Windows (VRPTW). The model considers the minimization of emissions and fuel consumption as the primary or secondary objective and proposed a solution approach for different levels of congestion. A heuristic was presented to help reduce emissions once some feasible routes are obtained for the time-dependent VRPTW. From the results, the author concludes that the effects of congestion on the emission levels are not uniform.

Urquhart et al. [23] examined the Traveling Salesman Problem (TSP) based on real-world data using vehicle emissions as the fitness criteria when using evolutionary algorithms (EA) to solve the routing problem. A low-cost path finding algorithm is used to build paths, and the EA is then used to discover tours using paths with low emissions characteristics. Two methods are used to estimate the $CO_2$ emissions and the results are compared with each other. They provided conclusions about the trade-off between $CO_2$ savings, distance, and the number of vehicles used.

Bektaş and Laporte [24] introduced the Pollution-Routing Problem (PRP) as an extension of the classical VRP but considering a broader objective function that not only takes into account the travel distance but the costs of carbon emissions along with those of labor and fuel. They expressed costs as a function of load, speed, and other parameters. The proposed model is solved by the branch-and-cut method, which seems to work for integer programming problems. They concluded that the traditional objective of distance minimization does not necessarily imply that fuel cost or labor is reduced as well. Also, a solution with minimal cost does not mean one with minimal energy.

Environmental costs are considered in the objective function of a vehicle routing problem in work presented by Faulin et al. [25]. Apart from the costs caused by polluting emissions, they considered the environmental costs derived from noise and congestion. Algorithms with Environmental Criteria (AWEC) were constructed to address the need for solutions to real problems in delivery companies in a case study in Spain. They highlighted the importance of designing an algorithm that not only optimizes the distribution costs but also the costs associated with environmental impact. The results show that the AWEC can find quick solutions in routing problems with less than 100 nodes. However, the inclusion of environmental costs in the problem increased optimization cost by around 28%.

The research conducted by Ubeda et al. [26] is the first to incorporate the minimization of greenhouse gas emissions in the Vehicle Routing Problem with Backhauls. Both distances and pollutant emissions are minimized. The results of their study revealed that backhauling seems to be effective in controlling emissions, so it could be implemented by companies to improve efficiency in energy and to help reduce the impact on the environment.

Suzuki [27] developed an approach to the time-constrained, multiple-stop, truck-routing problem that minimizes the fuel consumption and pollutant emission. The analysis showed a significant saving in fuel consumption when heavy items were delivered in the initial segments of a tour, and light items were delivered in the latter segments. In this way, the distance a vehicle travels with heavy payloads can be minimized. The latter produced fuel savings between 4.9% and 6.9% over the conventional shortest-distance approach, and between 1.0% and 3.1% over the standard minimal-fuel approach.

An extended Adaptive Large Neighborhood Search (ANLS) for the PRP was proposed by Demir et al. [28] to enhance the computation efficiency for medium or large-scale PRP. To evaluate the algorithm's effectiveness, they generated different sets of instances based on real geographic data. The results of the procedure were compared against the solutions obtained using the integer linear programming formulation of the PRP. The algorithm was effective and found quality solutions on instances with up to 200 nodes.

Jemai et al. [29] implemented the NSGA-II evolutionary algorithm to the bi-objective GVRP in the context of green logistics to minimize the total traveled distance and the total $CO_2$ emissions. They applied evolutionary algorithms to solve GVRP benchmarks and to find better Pareto fronts for the problem of study and performed a statistical test to confirm their quality. The results prove the effectiveness of considering the minimization of emissions as a separate objective.

Erdoğan and Miller-Hooks [30] conceptualized and formulated a GVRP as a mixed-integer linear program to help organizations with alternative fuel-powered vehicles to overcome difficulties due to limited driving range and limited infrastructure to refuel. To minimize the total distance, the model tries to reduce the risk of running out of fuel. The Modified Clarke and Wright Savings heuristic, the Density-Based Clustering Algorithm, and a customized improvement technique were developed. The results showed that the techniques perform well in comparison to exact solution methods, which can be used to solve large problem instances.

Xiao et al. [31] develop a formulation of fuel consumption. They presented a mathematical optimization model and proposed a Fuel Consumption Rate (FCR) to be added to the CVRP and extend it to minimize fuel consumption (FCVRP). In their research, both the traveled distance and the load are considered as factors to determine fuel costs. They also developed a Simulated Annealing (SA) algorithm with a hybrid exchange rule for the proposed model. The results show that in comparison with the CVRP, the proposed model can reduce fuel consumption by 5% on average.

Huang et al. [32] presented a GVRP with a simultaneous pickup and delivery problem (GVRPSPD). A linear integer programming model was developed, including $CO_2$ emission and fuel consumption costs, modified from the commodity flow-based VRPSPD formulation. The numerical results show that the proposed model generates more environment-friendly routes without impacting a lot on the total distance traveled.

Ramos et al. [33] proposed a decomposition solution method for the Multi-Product, Multi-Depot VRP (MP-MDVRP). The primary goal of the research is to define service areas and vehicle routes that minimize the $CO_2$ emissions of a logistic system with multiple products and depots. The proposed method is solved using the branch-and-bound algorithm.

A vehicle transportation and routing model for alternative fuel vehicles (AFV) was introduced by Omidvar and Tavakkoli-Moghaddam [34]. The model aims to minimize the energy and fuel consumption in which GHG (Green House Gases) emissions are primary objectives. Some strategies are carried out and modeled to avoid congestion during traffic peak hours. A Simulated Annealing (SA) and Genetic Algorithm (GA) are proposed along with a partial heuristic method and an exact algorithm for solving small-scale problems.

Li [35] presented a mathematical model for the VRP with Time Windows (VRPTW) with a new objective function of minimizing the total fuel consumption. A Tabu Search (TS) algorithm with a random variable neighborhood descent procedure (RVND) is performed to solve the problem. The results showed that the solution of the proposed model has the potential of saving fuel consumption contrary to the results of the traditional VRPTW.

Jabali et al. [36] introduced the Emissions-based Time-Dependent Vehicle Routing Problem (E-TDVRP). The model considers travel time, fuel, and $CO_2$ emissions costs and considers limiting vehicle speed as part of the optimization because of the correlation of the emissions with speed. A Tabu Search procedure is used to solve the problem. Using the results obtained, they showed that reducing emissions leads to reducing costs, and limiting vehicle speeds is desired from a total cost perspective.

Franceschetti et al. [37] described an integer linear programming formulation for the Time-Dependent Pollution-Routing Problem (TDPRP) that consists of routing a fleet of vehicles in order to serve a set of customers and determining the speeds on each leg of the routes. The cost function includes emissions and driver costs and takes into account traffic congestion since it restricts vehicle speeds and increases emissions at peak hours.

The Vehicle Routing and Scheduling Problem (VRSP) in picking up and delivering customer to airport service was introduced by Peiying et al. [38] as an extension of the classical VRP but with more resource constraints. They presented a mathematical model to minimize the cost based on low carbon emissions and developed a Bi-directional Optimization Heuristic Algorithm (BOHA) for this model. The results showed that the model and algorithm could provide data and help the airport decision-makers develop economical and punctual shuttle service that improves customer satisfaction.

Kwon et al. [39] presented the heterogeneous vehicle routing problem that determines a set of vehicle routes that satisfies customer demands and vehicle capacities and minimize the sum of variable operating costs. An integer programming model was used, and for more complex problems, Tabu Search algorithms were developed and found solutions within a reasonable computation time. The results showed that the amount of carbon emissions could reduce without increasing the total costs by carbon trading.

Yasin and Yu [40] developed a mathematical model and a Simulated Annealing (SA) heuristic for the Green Vehicle Routing Problem (GVRP). The objective of the GVRP is to minimize the total distance traveled by the alternative fuel vehicle fleet. Computational results indicate that the heuristic can obtain good GVRP solutions in a reasonable amount of time.

Küçükoğlu et al. [41] analyzed and formulated the Green Capacitated Vehicle Routing Problem (GCVRP), which aims to minimize total fuel consumption of a route and, consequently, the $CO_2$ emission of the vehicles in the route. A mixed-integer linear programming model is proposed for solving the problem. They computed fuel consumption considering the vehicle's technical specifications, vehicle load, and the distance. The results show that the GCVRP model provides significant reductions in fuel consumption and that the minimum distance not necessarily represents minimum consumption.

Pradenas et al. [42] studied the decrease in the emission of greenhouse gases using the VRP with backhauls and time windows by considering the energy required for each route and estimating the load and distance between customers. The scatter search algorithm that minimizes GHG emissions for a homogeneous vehicle fleet was designed and implemented. The results indicate that the distance traveled, and the transportation costs increase as the energy required and fuel consumed decrease. In consequence, the emission of greenhouse gases decreases too.

The Emission Minimization Vehicle Routing Problem with Vehicle Categories (EVRP-VC) was introduced by Kopfer et al. [43] to minimize fuel consumption and $CO_2$ emissions instead of driving distances. The fuel consumption depends on the weight of the vehicle during its route. The model also considers employing different types of vehicles in order to exploit their dead weights, maximal payloads, and fuel efficiency. The results showed the potential for a significant reduction. However, on the other hand, the total distances traveled and the number of routes of the used vehicles increase significantly, which may cause congestions.

Treitl et al. [44] proposed an integer programming model for the Inventory Routing Problem to minimize total transport costs and $CO_2$ emissions costs from transport and warehousing activities. They focused on analyzing transport processes and in the economic and environmental effects of routing decisions in a supply chain with vertical collaboration.

The Green Capacitated Vehicle Routing Problem (GCVRP) was analyzed by Úbeda et al. [45] as an extension of the classical CVRP with a minimizing environmental emissions objective. They based the objective on calculating $CO_2$ emissions, which are dependent on factors such as speed, weather, load, and distance. A Tabu Search (TS) approach is used to solve the problem. The results showed that appropriate fleet planning might balance total pollution and the total cost more effectively, and the TS approach adapts the environmental criterion better than other procedures.

Taha et al. [46] presented an integer programming based exact solution model for the small size GVRP instances. The problem aims to minimize the travel distance while reducing $CO_2$ emissions using alternative sources of fuel. The solution aims to help organizations that operate alternative fuel-powered vehicles to eliminate the risk of running out of fuel while sustaining low-cost routes.

Ayadi et al. [47] proposed a mathematical model for the GVRP with multiple trips (GVRPM) where vehicles are allowed to take more than one route during the working day. The objective is to minimize the amount of greenhouse gas emissions. An Evolutionary Algorithm is developed to solve the problem by combining a Genetic Algorithm with a Local Search procedure. The results showed that this method produced a better routing plan with lower emissions but longer transportation times and traveled distance.

Adiba et al. [48] presented a version of the classical Capacitated Vehicle Routing Problem (CVRP) to minimize the greenhouse gas emissions and the total traveled distance. A Genetic Algorithm was applied to solve the problem, and the results showed that this method led to calculate the emissions more accurately and facilitate the search for cleaner routes.

Montoya et al. [49] studied the GVRP as an extension of the classical VRP in which routes are performed using alternative fuel vehicles (AFV). The GVRP presented duration and renewable fuel consumption constraints. They proposed an effective route-first, split-second hybrid heuristic to solve the problem and introduced a novel mechanism to split giant tours into GVRP solutions. The approach was tested on a set of 52 instances from the literature.

Ene et al. [50] presented the GVRP with a heterogeneous fleet for both capacity and time windows constraints to reduce fuel consumption and minimize $CO_2$ emissions. They developed a Simulated Annealing and Tabu Search based hybrid metaheuristic algorithm to analyze the effect of a heterogeneous fleet on reducing fuel consumption. The vehicle technical data, speed, weight, and acceleration, along with the distance traveled, were considered to calculate the fuel consumption.

ÇağrıKoç and Karaoglan [51] proposed a Simulated Annealing (SA) heuristic-based exact solution approach to solve the GVRP considering a limited driving range of vehicles in conjunction with limited refueling infrastructure. The exact algorithm bases on the branch-and-cut algorithm, which combines several valid inequalities to improve lower bounds and a heuristic algorithm based on SA to obtain upper bounds. The results showed that the proposed approach was able to find high-quality solutions.

Afshar-Bakeshloo et al. [52] developed the Satisfactory-Green Vehicle Routing Problem (S-GVRP) consisting of routing a heterogeneous fleet of vehicles in order to serve a set of customers within predefined time windows. In addition to the objective of the classical VRP, both pollution and customer satisfaction were taken into account. The proposed mixed-integer formulation applied piecewise linear functions to a nonlinear fuzzy interval for incorporating customers' satisfaction into linear objectives. The model solved the problem to optimality.

Arango Gonzalez et al. [53] presented a problem for waste collection in islands in the south of Chile. The transportation is done by ships at docks (collection sites). Beyond the aspect of pollution caused by the transportation, also the environmental aspect is considered by the leaching and decomposition of the waste that contaminate the soil and attract wildlife and insects. Here, waiting time is important to minimize these environmental impacts. A periodic vehicle routing model was proposed to solve the bi-objective problem.

Andelmin and Bartolini [54] developed an exact algorithm for solving the GVRP based on a set partitioning formulation using a multigraph. The GVRP models the optimal routing of an alternative fuel vehicle fleet to serve a set of geographically scattered customers. They strengthen a set partitioning formulation by adding valid inequalities, including k-path cuts, and a method for separating them was described. The results showed that the exact algorithm could optimally solve instances with up to 110 customers.

More recently, the authors developed a Multi-Start Local Search heuristic for the GVRP [55], which iteratively constructs new solutions, stores the routes, and optimally combines these routes by solving a set partitioning problem. The heuristic was tested and the results were compared to other heuristics on a set of benchmark instances with up to 500 customers.

Yu et al. [56] proposed a Hybrid Vehicle Routing Problem (HVRP) as an extension of the GVRP. They generated a mathematical model to minimize the total cost of travel by driving Plug-in Hybrid Electric Vehicles (PHEV). The model considers the use of electric and fuel power depending on the availability of electric or fuel stations. A Simulated Annealing with a Restart Strategy was developed to solve the problem.

L. N. U. Cooray and Rupasinghe [57] implemented a Genetic Algorithm (GA) to solve the Energy-Minimizing Vehicle Routing Problem (EMVRP) and used Machine Learning Techniques to determine the parameters of the GA. In the EMVRP, the energy is generalized to be equal to a function of load and distance.

A research on multi-objective green logistics optimization was conducted by Sawik et al. [58]. Optimality criteria focused on minimizing environmental costs: the amount of money paid for noise, pollution, and costs of fuel versus noise, pollution, and fuel consumption themselves. They proposed and applied some mixed-integer programming formulations of multi-criteria Vehicle Routing Problems for finding optimal solutions.

Toro et al. [59] introduced the Green Capacitated Location Routing Problem (G-CLRP) as an extension of the CLRP considering fuel consumption minimization to include greenhouse effect costs related to the environmental impact of transportation activities. A mathematical model is developed as a mixed-integer linear problem with a bi-objective formulation considering the minimization of both operational costs and environmental effects. The sensitivity analysis results showed that the proposed objective functions were conflicting goals. An epsilon constraint technique was used to solve the proposed model.

The Green Open Location Routing Problem (G-OLRP) was introduced by Toro et al. [60]. The classic OLRP involves the selection of one or many depots from a set of candidates and the planning of delivery radial routes from the selected depots to a set of customers. The model proposed is formulated as a bi-objective problem that considers the minimization of both operational costs and environmental effects. An epsilon constraint technique was used to solve this model. The trade-off between economic aspects and environmental aspects can be selected from the Pareto front using a decision-making process.

Hamid Mirmohammadi et al. [61] developed a mixed-integer linear mathematical programming model for the Time-Dependent Periodic Green Vehicle Routing Problem that considered the time windows to serve a set of customers and multiple trips, assuming that urban traffic would disrupt timely services. The objective function aimed to minimize the $CO_2$ emissions, earliness, and lateness penalties costs and used vehicles' cost.

De Oliveira da Costa et al. [62] proposed a Genetic Algorithm (GA) to address the GVRP, aiming to minimize the $CO_2$ emissions of the routes. With this objective, not only greenhouse gases could be reduced, but resulting routes may be financially cost-effective. The GA solution approach incorporated elements of Local and Population Search Heuristics. The solutions obtained were compared with routes used by drivers in a courier company and proved that reduction in emissions was achieved without additional operational costs.

Tirkolaee et al. [63] presented a model for the Multi-trip Green Capacitated Arc Routing Problem (G-CARP) intending to minimize total cost of generation and emission of greenhouse gases, the cost of vehicle usage and routing cost. A Hybrid Genetic Algorithm was developed to solve the problem using a Simulated Annealing procedure to generate the initial solutions and a Genetic Algorithm to find the best possible solution. The proposed algorithm presented good performance in a suitable computational time.

Recently, a Green Mixed Fleet Vehicle Routing Problem with partial battery recharging and time windows has been introduced by Macrina et al. [64] as a variant of the GVRP with time windows. They used a mathematical model and an Iterated Local Search metaheuristic to solve the problem. The possibility of partial recharge in electric vehicles and a limitation on the polluting emissions for conventional vehicles were considered. The results demonstrated the influence of the main features of the problem on solution configuration and costs. They showed that the use of partial recharge led to more effective and sustainable solutions.

In their recently published research, Wang and Lu [65] proposed a memetic algorithm with competition to solve the Capacitated Green Vehicle Routing Problem (CGVRP). They designed a particular decoding method, population initialization, multiple search operators on TSP route and CGVRP route, competitive search, and crossover operator. Extensive comparative results showed that the proposed algorithm is more effective and efficient than the existing methods to solve the CGVRP.

Li et al. [66] developed a Multi-Depot Green Vehicle Routing Problem (MDGVRP) with conflicting objectives: maximize revenue and minimize costs, time and emissions. They applied an Improved Ant Colony Optimization (IACO) to solve the problem that uses an innovative approach in updating the pheromone that results in better solutions.

Yu et al. [67] have recently addressed a Heterogeneous Fleet Green Vehicle Routing Problem with Time Windows (HFGVRPTW). They proposed an improved branch-and-price algorithm to solve the problem and developed a Multi-Vehicle Approximate Dynamic Programming (MVADP) algorithm to speed up the solution of the pricing problem. A tighter upper bound was quickly obtained by an integer branch method.

Recently, Dukkanci et al. [68] introduced the Green Location-Routing Problem (GLRP) as a combination of the classical Location-Routing Problem (LRP) and the Pollution-Routing Problem (PRP). The objective of the GLRP is to minimize a cost function that includes the fixed cost of operating depots and the costs of the fuel and $CO_2$ emissions. A mixed-integer programming formulation is developed to solve the problem describing some preprocessing rules and valid inequalities to strengthen the formulation. Two heuristic algorithms to solve small and larger instances are also proposed.

## 4.3  Problem Description

This section describes the general characteristics of the Cumulative Capacitated Vehicle Routing Problem (CCVRP) and the peculiarities that arise when integrating the priority factor in which the customers should be served.

We describe the Cumulative Capacitated Vehicle Routing Problem as follows: There is a known quantity of vehicles, each with different capacities, to service to a known number of clients (nodes), each with different demands. All vehicles depart from a central depot to their routes. Total demand served by the route of each vehicle must not exceed its capacity. A priority index in the service is introduced to the problem. This index affects the visiting sequence to the nodes on different routes. Based on the above, we establish the following characteristics of the problem:

- There is a known (fixed) number of vehicles to provide service, each with different known capacity.
- There is a known number of clients to be served, each with known demand.
- The number of customers attended by each vehicle will depend on its capacity.
- There is a common depot which sets the starting point of the route.
- A different route for each vehicle is set.
- Each route considers the course of each vehicle from the depot to the last customer to serve (the way back to the depot is not taken into account).
- All customers must be served.
- Every customer should be served by a single route (a vehicle).
- Each client should be served, preferably according to a designated priority factor for each of them.
- The designated order attention for clients must meet two main objectives: the priority of each client and the route latency.

In summary, the problem can be formally defined as an undirected graph $G = (V, A)$ where $V = \{0, 1, 2, \ldots, n\}$ denote the node set, and $A$ is the set containing all arcs. Node 0 corresponds to the depot, and the rest of the nodes form the set of customers $V' = \{1, 2, \ldots, n\}$. Each arc $(i, j) \in A$ has an associated travel time $c_{ij}$ and each node $j \in V'$ has an associated demand $d_j$. A heterogeneous fleet of $k$ vehicles is available, each with capacity $Q^k$. In addition, a preference matrix $P$ is defined in which $p_{ij} = 1$ represents that customer $i$ might be served before customer $j$ and $p_{ij} = 0$ means that customer $i$ can be served after the customer $j$.

To clarify the explanation, we define a tour as an ordered subset of nodes of $V$ with node 0 in the first position and it is represented by $P = \{0, [1], [2], \cdots, [r]\}$, where $m$ is the amount of nodes in the tour and $[i]$ means the node that occupies the position $i$ in tour $P$. Then, the latency of tour $P$ can be calculated as in (4.1) (see [69]):

$$Lat(P) = rc_{0[1]} + (r-1)c_{[1][2]} + (r-2)c_{[2][3]} + \cdots +$$

$$2c_{[r-2][r-1]} + c_{[r-1][r]} = \sum_{h=0}^{r-1}(r-i)c_{[h][h+1]}, \qquad (4.1)$$

where the elements $c_{ij}$ of the time matrix $C$ denote the travel time between a pair of customers. The latency of a given customer represents its waiting time to start the service. Since the structure of the problem corresponds to an open VRP, we do not include the travel time of the last customer $[r]$ on the path $P$ in the total latency calculation.

The goal is to find $k$ tours in such a way that $P_i \cap P_j = \{0\}$, $(i, j = 1, 2, \cdots, k; j \neq i)$, $\cup_{i=1}^{k} P_i = V$ and to minimize the function $L = \sum_{i=1}^{k} Lat(P_i)$, i.e., to find $k$ tours that have in common only node 0 in the first position, that together cover all of the nodes and minimize the sum of the latencies of the tours by satisfying the demand of all customers without exceeding the capacity of each vehicle. All customers should be served according to their priority level (preferably) and minimizing the system average latency (average waiting time of customers).

Also, for each path, the tardiness value arises when the arrival time $(t_i)$ for a customer of less priority index is minor than the arrival time of a customer with a higher priority index $(t_j)$ (even in different routes, $p_{ij} = 1$). We observed that qualitatively, the amount of tardiness is associated with customer satisfaction. However, because the first objective is estimated as a function of distance, to standardize satisfaction level calculation, the difference between their arrival times is calculated as $I_{ij} = t_j - t_i$. The total tardiness of the system is computed, as shown in (4.2):

$$\text{Total tardiness} = \sum_{i \in V'} \sum_{j \in V'} I_{ij}. \qquad (4.2)$$

## 4.4   Mathematical Formulation

For the formulation, the following parameters and variables are considered:

**Parameters**
- $N$: Number of customers
- $K$: Number of vehicles
- $d_j$: Demand for customer $j$
- $Q^k$: Capacity of vehicle $k$
- $Q_{max}$: Maximum capacity of any of the vehicles
- $M$: Maximum travel distance allowed (the same for all vehicles)
- $C$: Traveling cost matrix between every pair of nodes (including the depot)
- $P$: Preference matrix between every pair of nodes

**Variables**

$$w_{0j}^k = \begin{cases} 1, & \text{if the vehicle } k \text{ is assigned to a path from node 0 to customer } j \\ 0, & \text{otherwise.} \end{cases}$$

$$x_{ij} = \begin{cases} 1, & \text{if the arc } (i, j) \text{ is in the path of a vehicle} \\ 0, & \text{otherwise.} \end{cases}$$

$y_{ij} = $ number of remaining nodes after the node $i$ on a route if $x_{ij} = 1$;
otherwise 0

$v_{0j}^k = $ the sum of demands of the nodes after node 0 on the route k if $w_{0j}^k = 1$;
otherwise 0

$v_{ij} = $ sum of demands of the nodes after node $i$ on a route if $x_{ij} = 1$; otherwise 0

$t_{0j}^k = $ Arrival time of node $j$ from node 0 on a route $k$ if $w_{0j}^k = 1$; otherwise 0

$t_{ij} = $ Cumulative time at node $j$ on a route if $x_{ij} = 1$; otherwise 0

$l_{ij} = $ Tardiness in the arrival time to node $i$ if node $j$ is served first($p_{ij} = 1$);
otherwise 0

The corresponding formulation is stated as follows:
Minimize:

$$\sum_{i \in V'} c_{0i} y_{0i} + \sum_{\substack{i \in V' \\ }} \sum_{\substack{j \in V' \\ j \neq i}} c_{ij} y_{ij} + \sum_{\substack{i \in V' \\ }} \sum_{\substack{j \in V' \\ j \neq i}} l_{ij}, \tag{4.3}$$

subject to:

$$\sum_{j \in V'} \sum_{k \in K} w_{0j}^k = K, \tag{4.4}$$

$$\sum_{i \in V'} x_{i0} = K, \tag{4.5}$$

$$\sum_{j \in V'} \sum_{k \in K} w_{0j}^k = 1, \qquad \forall k \in K \tag{4.6}$$

$$\sum_{k \in K} w_{0j}^k + \sum_{i \in V'} x_{ij} = 1, \qquad \forall j \in V' \tag{4.7}$$

$$\sum_{j \in V} x_{ij} = 1, \qquad \forall i \in V' \tag{4.8}$$

$$y_{0j} + \sum_{\substack{i \in V' \\ i \neq j}} y_{ij} - \sum_{\substack{i \in V' \\ i \neq j}} y_{ji} = 1, \qquad \forall j \in V' \tag{4.9}$$

$$y_{0j} \geq \sum_{k \in K} w_{0j}^k, \qquad \forall j \in V' \tag{4.10}$$

$$y_{ij} \geq x_{ij}, \qquad\qquad\qquad\qquad \forall i \in V'; \forall j \in V'; i \neq j \qquad (4.11)$$

$$y_{0j} \leq (N - K + 1) \sum_{k \in K} w_{0j}^k, \qquad\qquad\qquad \forall j \in V' \qquad (4.12)$$

$$y_{ij} \leq (N - K) \sum_{k \in K} x_{ij}, \qquad\qquad\qquad \forall i \in V'; \forall j \in V' \qquad (4.13)$$

$$v_{0j}^k \geq d_j w_{0j}^k, \qquad\qquad\qquad\qquad \forall j \in V'; \forall k \in K, \qquad (4.14)$$

$$v_{0j}^k \leq Q^k w_{0j}^K, \qquad\qquad\qquad\qquad \forall j \in V'; k \in K, \qquad (4.15)$$

$$v_{ij} \geq d_j x_{ij}, \qquad\qquad\qquad\qquad \forall i \in V'; j \in V'; i \neq j \qquad (4.16)$$

$$v_{ij} \leq (Q_{\max} - d_i) x_{ij}, \qquad\qquad\qquad \forall i \in V'; j \in V; i \neq j \qquad (4.17)$$

$$\sum_{\substack{k \in K}} v_{0j}^k + \sum_{\substack{i \in V' \\ i \neq j}} v_{ij} - \sum_{\substack{i \in V \cap \{n+1\} \\ i \neq j}} v_{ji} = d_j, \qquad\qquad \forall j \in V', \qquad (4.18)$$

$$t_{0j}^k = c_{0j} w_{0j}^k, \qquad\qquad\qquad\qquad \forall j \in V'; j \in V'; k \in K, \qquad (4.19)$$

$$t_{0j}^k \leq M w_{0j}^k, \qquad\qquad\qquad\qquad \forall i \in V'; j \in V'; i \neq j; k \in K, \qquad (4.20)$$

$$t_{ij} \geq c_{ij} x_{ij}, \qquad\qquad\qquad\qquad \forall i \in V'; j \in V'; i \neq j$$

$$(4.21)$$

$$t_{ij} \leq (M - c_{ij}) x_{ij}, \qquad\qquad\qquad \forall i \in V'; j \in V; i \neq j$$
$$(4.22)$$

$$\sum_{\substack{h \in V \cap \{n+1\} \\ h \neq i}} t_{ih} - \sum_{\substack{h \in V' \\ h \neq i}} t_{hi} - \sum_{k \in K} t_{0i}^k = \sum_{\substack{j \in V \\ j \neq i}} c_{ij} x_{ij}, \qquad\qquad \forall i \in V',$$

$$(4.23)$$

$$I_{ij} \geq p_{ij} \left( \sum_{k \in K} t_{0i}^k + \sum_{\substack{h \in V' \\ h \neq i}} t_{hi} - \sum_{\substack{h \in V' \\ h \neq j}} t_{hj} - \sum_{k \in K} t_{0j}^k \right) \quad \forall i \in V'; j \in V'; i \neq j$$

$$(4.24)$$

$$w_{0j}^k \in \{0, 1\}, \qquad\qquad\qquad\qquad \forall j \in V'; \forall k \in K,$$
$$(4.25)$$

$$x_{ij} \in \{0, 1\}, \qquad\qquad\qquad\qquad \forall i \in V'; \forall j \in V,$$
$$(4.26)$$

$$y_{ij} \geq 0, \qquad\qquad \forall i \in V; \forall j \in V, \tag{4.27}$$

$$v_{0j}^{k} \geq 0, \qquad\qquad \forall j \in V'; k \in K, \tag{4.28}$$

$$v_{ij} \geq 0, \qquad\qquad \forall i \in V'; j \in V. \tag{4.29}$$

$$t_{0j}^{k} \geq 0, \qquad\qquad \forall j \in V'; k \in K, \tag{4.30}$$

$$t_{ij} \geq 0, \qquad\qquad \forall i \in V'; j \in V. \tag{4.31}$$

$$I_{ij} \geq 0, \qquad\qquad \forall i \in V'; j \in V'. \tag{4.32}$$

In this formulation, the objective function (4.3) aims searching a trade-off between the objective of the total latency of the system and the total tardiness of the system (when their priorities are not respected). Constraints (4.4) and (4.5) ensure that exactly $K$ arcs leave and return to the depot. Constraint (4.6) establishes that a customer node can be visited for exactly one vehicle coming from the depot. Constraints (4.7) and (4.8) impose that exactly one arc enters and leaves each node associated with a customer. Constraints (4.10)–(4.13) help to avoid subtours and allow to calculate a customer position on the routes. Constraints (4.12) force variables $y_{0j}$ to be zero when $w_{0j}^{k} = 0$ and constraints (4.13) force variables $y_{ij}$ to be zero when $x_{ij} = 0$. Constraints (4.14) and (4.15) are derived from a generalization of restrictions (4.10) and (4.11). Constraints (4.16) force variables $v_{0j}^{k}$ to be zero when $w_{0j}^{k} = 0$. Constraints (4.17) force variables $v_{ij}$ to be zero when $x_{ij} = 0$. Constraints (4.18) ensure that the demand at each node $i$ is fulfilled and in conjunction with (4.16) and (4.17) estimate the load per vehicle. Constraints (4.19)–(4.23) control the arrival time to the nodes (customers). Constraints (4.24) estimate the tardiness (in case of violating the priority constraints). Finally, constraints (4.25)–(4.32) establish the nature of the variables.

### 4.4.1 Reformulation Using Epsilon Constraint

In this subsection, we present an analysis of the characteristics of the proposed model. An essential task in multi-objective optimization is to find Pareto-optimal solutions. As a bi-objective approach, we decided to implement solution method in order to generate a good front of true efficient solutions.

In this work, we can note that the objective functions are separable; that is, each of them involves different decision variables. On the one hand, variables $y_{ij}$ allow calculating the total arrival time to the customers. On the other hand, variables $I_{ij}$ quantify the total tardiness for the case in which customers with a minor priority are served earlier than customers with a higher priority index.

In order to clarify this, the particular structure of the bi-objective problem herein proposed is further described:

$$\min_{y \in Y} F_1(y) = L = \sum_{i \in V'} c_{0i} y_{0i} + \sum_{i \in V'} \sum_{\substack{j \in V' \\ j \neq i}} c_{ij} y_{ij} \tag{4.33}$$

$$\min_{i \in I,} F_2(I) = T = \sum_{i \in V'} \sum_{\substack{j \in V' \\ j \neq i}} I_{ij} \tag{4.34}$$

subject to :

$$(4.4)-(4.24).$$

The mentioned characteristic of the bi-objective model is exploited in the proposed method. Hence, we propose the $\varepsilon$-constraint method as a solution procedure. For every single routing decision (4.4)–(4.18), the minimum tardiness (min $T$) problem bounded by constraints (4.19)–(4.24) is solved as principal objective function, transforming the latency function $(L)$ into a constraint. In order to efficiently implement the method to find solutions belonging to the Pareto front, a variant of the $\varepsilon$-method, known as AUGMECON [70], was implemented.

## 4.5   Metaheuristic Algorithm

This section describes the metaheuristic approaches implemented to solve the problem effectively. Both procedures are developed based on population procedures since they have shown their effectiveness in solving sizeable combinatorial optimization problems.

### 4.5.1   Non-dominated Sorting Genetic Algorithm

Holland et al. [71] first proposed Genetic Algorithms (GAs) inspired on evolution theory ideas. Due to their simple and yet effective search procedure, several papers such as [72] describe their successful implementations in vehicle routing problems. Srinivas and Deb [73] introduced a multi-objective version of this algorithm, and Deb et al. [74] improved it under the name of Non-dominated Sorting Genetic Algorithm (NSGA). In general, the NSGA implemented in this work is based

---

**Algorithm 1:** Non-dominated Sorting Genetic Algorithm

---

1 **begin**
2    $it \leftarrow 0$;
3    Initialize a population $(P_0)$ of $\sigma$ chromosomes;
4    Sort $P_0$ in fronts following the non-dominated sorting approach;
5    **for** $(it = 1; it <= Maxiter; it++)$ **do**
6       Generate an offspring population $P_t$, of size $N$, from $P_{it-1}$, using selection,
         crossover and mutation operators.;
7       Combine parent and offspring population $R_t = P_t \cap P_{t-1}$.;
8       $u \leftarrow 0$ ;
9       **repeat**
10          Select randomly and individual $E$ from $P_{t-1}$;
11          Perform a mutation operation over $E$;
12          Include the individual into $P_t$;
13          Sort $P_t$ in fronts following a non-dominated sorting approach;
14          $u \leftarrow u + 1$;
15       **until** $u = D$;
16       Sort population using the non-dominated sorting approach and identify fronts
         $F_i = (1, 2, \cdots)$ and calculate the crowding distance for each solution in $F_i$;
17       Make $T_{it+1} \leftarrow \emptyset, i \leftarrow 1$;
18       **while** $(|T_{it+1}| + |F_i| < N)$ **do**
19          $T_{it+1} \leftarrow T_{it+1} + F_i$ ;
20          $i \leftarrow i + 1$;
21          Sort solutions in $F_i$ in decreasing order accordance with crowding distances,
         select the first $N - |T_{t+1}|$ elements of $F_i$ and add it to $T_{it+1}$;

---

**Fig. 4.1** Overall procedure of the proposed NSGA

on a version successfully implemented in [75] and consists of the following
mechanisms (operators): initialization, recombination (crossover and mutation), and
classification in fronts. The constructive procedure creates an initial population
of feasible solutions of size $N$. After that, during a predetermined number of
successive generations (iterations), the next population $(P_t)$ of size $N$ is generated
by implementing mechanisms of recombination and mutation between members
that represent tentative solutions (high-quality or non-dominated solutions) and
members that represent diverse solutions. Then, the individuals belonging to the
previous and current generations are evaluated and grouped in fronts, according
to the level of dominance. After this, a new population of size N is created by
inserting the individuals to the set, starting from the one belonging to the front of
non-dominated solutions. Figure 4.1 shows the pseudocode of the overall NSGA
procedure.

#### 4.5.1.1 Constructive Procedure

The constructive procedure creates an initial population of feasible solutions based on generating a chain $S_a = \{1, 2, \cdots n\}$ and, for each customer, an auxiliary random key $R_a$ is used to encode the solution by assigning a real number drawn randomly from [0, 1) to every single position. Also, an empty set $S_p$ is used to save the temporary assignment of customers to routes. After this, a simple decoding mechanism is applied over $R_a$ to insert the customers in $S_a$ into a temporary set $S_p$ (procuring to maintain feasibility in the capacity of the vehicles). The set $S$ represents a feasible initial solution of routes. The customers already inserted in $S$ are removed from $S_a$.

The iterative assignment mechanism operates as follows: First, the procedure initializes the first position in every route of the set $S_p$ as 0. Then, it sorts the customers in $S_a$ in a non-increasing order based on the information in $R_a$. Then, the decoding mechanism is applied as follows: in every iteration, the algorithm selects the n-th customer in $S_a$ and systematically tries to insert it into the first available position. For instance, if in the first possible route, there are two customers already assigned, the next open position will be the third one. It is important to emphasize that the construction procedure builds the routes in parallel. In other words, it performs the evaluation of feasible insertions over all of the routes. If so, the algorithm continues by selecting the next customer at $S_a$. The algorithm finishes when all customers have been added into $S_p$ or when the next customer to evaluate cannot be inserted.

If the algorithm reaches a feasible assignment, then $S \leftarrow S_p$ and is inserted to the population $Q_t$. Otherwise, the algorithm destroys the partially constructed solution in $S_p$ and generates a new random chain $S_a$ (with the auxiliary vector $R_a$).

After reaching a feasible assignment, routes are constructed, respecting the sequence in which they were assigned. Then, its corresponding objective functions $L$ (representing the total latency of the system) and the calculation of the tardiness $T$ based on the priorities of the customers) are performed, respectively. Figure 4.2 shows the pseudocode for this algorithm.

#### 4.5.1.2 Crossover Procedure/Mechanism

We base the crossover mechanism on combining two solutions $A$ and $B$ to create a new solution $C$. For this procedure, a crossover procedure that incorporates a tournament selection operator is performed (Fig. 4.3).

For this, two individuals from the current generation $P_t$ are selected. The first individual belongs to the front $F_0$, whereas the second is randomly selected from the entire population (generation). Subsequently, the customers of each solution are grouped into a single chain following the order they entered into the routes (the chain is constructed from the customer of the first route to the last customer of the latter route). As a result, the chains corresponding to chromosomes $A$ and $B$ are obtained. After this, a probability for inheriting is assigned to each solution. Those

---

**Algorithm 2:** Constructive procedure $(S, L, D)$

```
1 begin
      Data: S ← ∅, S_p ← ∅, S_a =← ∅, L ← 0, T ← 0
2     while (S_a ≠ ∅) do
3         Fill chain in S_a = {1, 2, ⋯ , n};
4         Create an auxiliary random key chain R_a;
5         Sort customers in S_a in a non-decreasing order according with their
            corresponding random values in R_a;
6         i = 0;
7         flag = 0;
8         while S_a ≠ ∅ or flag = 1 do
9             Select the corresponding customer i from S_a;
10            l = 0;
11            if Customer i is feasible to insert in route l then
12                Insert customer in the l − th route S_p;
13                Remove customer i from S_a;
14                i + +;
15                flag = 0;
16            else if h < K then
17                h + +;
18            else
19                Destroy the partial solution in S_p.;
20                Establish S_a = {1, 2, ⋯ n};
21                Create a new random key R_a ;
22                flag = 1 ;
23        Compute the values of total latency L and tardiness T for the individual;
```

---

**Fig. 4.2** Pseudocode of the constructive procedure

probabilities are complementary. If the first individual has a probability of $\alpha$, then the second individual will receive a probability of $1 - \alpha$. To enhance the creation of good quality solutions, the probability assigned to the first parent is always greater than 50%.

For every single position to fill (in the new solution), the roulette wheel is spin to determine if the element belonging to the first or second individual (chromosome) must be selected. If during the evaluation, one of the customers to choose has been previously assigned, the algorithm immediately evaluates if the other element is available to insert. If so, it is inserted. In case both customers of that position have been already assigned, then the algorithm continues evaluating the next position. If, after the evaluation of all positions unassigned customers still exist, the algorithm will try to place them in subsequently available positions in the child following the lexicographic order.

Once a child solution $C$ is obtained, its feasibility is evaluated by applying the constructive procedure. If feasible, their total latency $L_C$ and Tardiness $T_C$ values are computed. On the contrary, the child is discarded, and a new second parent is selected (preserving the first parent) from $P_t$ to restart the crossover procedure.

---

**Algorithm 3:** Crossover procedure $(S, L, D)$

---

1  **begin**
    **Data:** $Q, R, U$
2      **for** *(h = 1; h <= D; h + +)* **do**
3          Select randomly two parents $(A, B)$ from $Q$;
4          $i = 0$;
5          **repeat**
6              Spin the wheel and obtain the value of *probability*;
7              **if** *probability ≤ β and customer in A is available* **then**
8                  Select the customer of the chromosome A;
9              **else if** *probability ≤ β and customer in B is available* **then**
10                 Select the customer of chromosome B;
11             **else**
12                 i++;
13         **until** *All positions in both parents have been evaluated*;
14         **if** *unassigned customers remain* **then**
15             Assign them into the remaining available positions by following the lexicographic order;
16     Construct a feasible solution for the new individual;

---

**Fig. 4.3** Pseudocode of the crossover procedure

### 4.5.1.3   Mutation Procedure/Mechanism

The mutation mechanism is based on local search procedures. Moves of interchange and reallocation are employed over intra-route and inter-route procedures in order to intensify the search in pursuit of finding local minima. This mechanism has been successfully applied for a mono-objective version of the CCVRP [10]. Below we describe each type of move:

- *Intra-route interchange.* Given two customers belonging to the same route, their positions are interchanged into the route. If the customers belong to positions $h$ and $i$, arcs $(h − 1, h)$, $(h, h + 1)$, $(i − 1, i)$, and $(i, i + 1)$ are replaced by arcs $(h − 1, i)$, $(i, h + 1)$, $(i − 1, h)$, and $(h, i + 1)$. It is important to remark that these movements must always be feasible in terms of capacity.
- *Intra-route reallocation.* For a given customer its best position inside of the route is identified. If the best-identified position is different from the current one, the movement is performed.
- *Inter-routes interchange.* Two customers belonging to different routes (one form each route) will be interchanged as long as feasibility is preserved (in terms of capacity).
- *Inter-routes reallocation.* For a given customer its best position is identified in any of the routes. If the best-identified position is different from the current one, the movement is performed.

---

**Algorithm 4:** Local Search($S$, $L$, $T$)

1 **begin**
2      **repeat**
3          $S^* = S$ and $L^* = L$, and $T^* = T$;
4          S',L' T' ← Intra-route local search (S,L, T);
5          **if** *The solution is non-dominated* **then**
6             $S = S'$ and $L = L'$, $T^* = T$;
7          S'',L'' T''← Inter-routes local search (S,L, T);
8          **if** *The solution is non dominated* **then**
9             $S = S''$ and $L = L''$, $T^* = T$;
10      **until** $L^* > L$;
11      **return** $S^*, L^*, T^*$;

---

**Fig. 4.4** Mutation procedure

Both types of local searches are applied iteratively. Initial solutions are improved by applying the intra-local search procedure first. Next, the obtained local minimum is improved by applying the inter-routes local search. These procedures are iteratively applied, while the current solution value $L$ keeps improving. In each procedure, the reallocation movement is performed first, and the execution of the interchange movement next. The first improvement criterion (*FI*) is used. Figure 4.4 exhibits this process.

The mutation procedure was only applied to a certain percentage of individuals to accelerate the performance of the algorithm (in terms of CPU time). To determine which individuals shall be mutated, the following procedure is defined: For each individual, a real number drawn randomly from $[0, 1)$ is assigned. If this value is less than or equal to a threshold value of $\beta$, the mutation procedure is applied; otherwise, the chromosome remains without modification.

As observed, this mutation procedure seeks to insert improved individuals to the next generation. It is important to remark that the mechanism does not guarantee that the chromosome selected can be profoundly improved. Because the size of this subset is relatively small, it is always possible to find a chromosome to be improved.

## 4.5.2 Particle Swarm Optimization Algorithm

Kennedy and Eberhart [76] proposed Particle Swarm Optimization as a method to solve hard optimization problems for continuous variables. Its procedure is based on the interaction between particles (solutions) after some iterations. These particles form the swarm, mimicking biological behaviors like those of bird flocks and fish schools. There is a swarm intelligence based on stored collective knowledge about a "best" direction to move all the swarm particles. At the same time, there is an inertia of each particle that resists the swarm direction, based on its knowledge of the "best"

---

**Algorithm 5:** Constructive procedure for PSO $(K, V', SwarmSize, SW)$

---

1  **begin**
       **Data:** $SW \leftarrow \emptyset$
2      **for** $(s = 1; s <= SwarmSize; s++)$ **do**
3          Create a list $U \subseteq V'$ with unassigned customers;
4          **for** $(k = 1; k <= K; k++)$ **do**
5              Assign randomly an unassigned customer $i$ to route $k$;
6              Update $U \leftarrow U \setminus \{i\}$;

7          **for** $(i \in U)$ **do**
8              Select customer $i$ randomly;
9              Sort current routes by number of customers assigned, in non-decreasing order;
10             Assign customer $i$ to the first route in the list with enough available capacity;
11             **if** *(customer i canot be assigned to any route)* **then**
12                 Destroy the current solution and rebuild the solution;
13             Update $U \leftarrow U \setminus \{i\}$;

14         Calculate Latency $L_s$ and Tardiness $T_s$;
15         Add particle $s$ to $SW$, $SW \leftarrow SW \cup \{s\}$;

---

**Fig. 4.5** Pseudocode of the constructive procedure of PSO

direction. Combining of these directions results in an extensive exploration of the neighborhood and the search space that allows escaping from local optimums. The balance between the individual knowledge and the swarm knowledge determines the convergence speed to an approximate global optimum [77–79].

Although some continuous encoding mechanisms have been proposed for VRP problems [80], the method proposed in this work uses movements of customers between routes according to the memory of the best routes assigned.

The algorithm developed creates initial solutions (particles), and a certain number of iterations repeats the swarm movements. The guiding function for the movement is an aggregated function that combines Latency $(L)$ and Tardiness $(T)$, i.e., $\delta * L + (1 - \delta) * T$. The algorithm uses a fixed value of $\delta$, and the solutions obtained in the last iteration are filtered to keep only the non-dominated solutions.

Figure 4.5 shows the constructive procedure for the initial solutions. In this procedure, the customers are assigned to routes trying to construct feasible solutions as fast as possible. Here, $SW$ is the set of solutions (particles) in the swarm. The movement procedure is shown in Fig. 4.6. Here, the best global particle is identified in the swarm in each iteration. For each other particle in the swarm, a perturbation tries to move its customers to the routes where they are assigned in the best global particle. The perturbation is designed to maintain feasibility moving first the customers with the lowest demand. The first feasible movement is accepted, and a new particle is perturbed.

---

**Algorithm 6:**  Movement procedure for PSO $(SW, \beta)$

---

1 **begin**
2    **for** *(s ∈ SW)* **do**
3       ⌊ Calculate $Z_s = \beta * L_s + (1 - \beta) * T_s$
4    Sort particles $s \in SW$ in non-decreasing order according to the values of $Z_s$;
5    Find $s'$ with minimum $Z'_s$;
6    **for** *(s ∈ SW \ {s'})* **do**
7       $MC \leftarrow \emptyset$;
8       **for** *(customer i ∈ V'_s)* **do**
9          ⌊ Identify routes $k_{is}$ and $k_{is'}$ where customer $i$ is assigned in the current particle $s$ and in the best particle $s'$;
10       Find customers $i$ with different $k_{is}$ and $k_{is'}$, $MC \leftarrow MC \cup \{i\}$;
11       Sort customers $i \in MC$ in non-decreasing order, by demand $d_i$;
12       **for** *(i ∈ MC)* **do**
13          Exchange customer $i$ with any other customer $j$ in route $k_{is'}$ in its current particle if capacity is not exceeded;
14          If no exchange is feasible, exit, and try with next $i$;
15          If the exchange was feasible, exit, and modify a new particle $s$;

---

**Fig. 4.6** Pseudocode of the movement procedure of PSO

## 4.6 Computational Experience

### 4.6.1 Set of Instances and Parameters Tunning

This section reports and discusses the computational results for both algorithms over a set of benchmark instances. The instances used to conduct the experimentation were adapted from the ones obtained in the literature: Koulaeian et al. [81] (Kou15), Chunyu and Xiaobo [82] (CaX10), Gillett and Johnson [83] (GaJ76-7 to GaJ76-12), and Augerat et al. [84] (Pn16k8 and Pn23k8). The generated problem instances are characterized by the following criteria (i) number of customers, (ii) number of vehicles, (iii) coordinates (x,y) for all locations (including the depot), (iv) demand of each node, and (v) priority index for each node. The size of the instances ranges between 12 and 100 nodes and from 2 to 10 vehicles. Since the original instances do not consider any preference index between the customers to be served, we include a priority parameter by assigning a numerical index between $\{1, \frac{n}{5}\}$ based on a uniform distribution. The customers having the highest value of the index represent the ones that should be first served (highest level of priority). Given the modification of these instances to include the priority indexes, and to facilitate the reporting of results, they were renamed using the nomenclature "NOGC-x" where $x$ represents a consecutive number assigned according to the rank assigned to the instance (in terms of the number of nodes), by following a non-increasing order. For example, the instance with the lowest number of nodes is based on Kou15. Therefore, it was renamed as NOGC1. The remaining instances based on Pn16k8, CaX10, Pn23k8,

GaJ76-7, GaJ76-8, GaJ76-9, GaJ76-11, and GaJ76-12 were renamed as NOGC2, NOGC3, NOGC4, NOGC5, NOGC6, NOGC7, NOGC8, and NOGC9, respectively. In the case of the instance GaJ76-10, it was named NOGC10 because it has the largest size in the number of customers.

In the case of the NSGA, the values for the parameters corresponding to the size of the population $N$, the threshold value $\beta$, and the maximum number of generations $D$ are set as 1000, 0.1, and 100, respectively. For the PSO algorithm, the swarm size, number of iterations, and $\delta$ were set as 250, 50, and 0.5, respectively. We performed several tests with different number of iterations, and we found that, for all the analyzed instances, the PSO algorithm stops improving the solution some iterations before the iteration 50, depending on the instance. Both algorithms were executed 10 times, and we report the best execution for each instance.

All of the experiments were performed using a PC Intel Core i7 @2.30 GHz with 16 GB of RAM Memory under Windows 10 as OS. The formulation was modeled using Visual Studio and solved using CPLEX 12.8 under CONCERT technology. For each instance, we set a time limit of 7200 s (2 h). In the case of the GA and PSO algorithms, these were coded using the C++ language. The next sections report the numerical results obtained by each algorithm over the test instances.

## 4.6.2   Experimental Results for the Solved Instances

The first set of experiments aims to evaluate the accuracy of the heuristic concerning optimality (optimally solved instances), comparing the results with those obtained by the resolution of the model. We first present the results for the instances NOGC1 and NOGC2 (up to 15 nodes). In particular, for the instance NOGC2, in the last six iterations of the AUGMECON, the solver reached the time limit, and the best-integer solution obtained is reported.

We use the following metrics to compare the performance of the exact and approximation procedures:

- Number of points on the front. The larger, the better.
- CPU time (in seconds). The smaller, the better.
- $k$-distance [85]. The smaller, the better.
- Hypervolume [86]. The larger, the better.
- The coverage of the fronts [86] computed of one front over another, denoted by $c(X', X'')$. The higher, the better.

These metrics have been successfully applied in bi-objective VRPs [75].

The number of non-dominated points measures the ability of each method to find efficient fronts. Table 4.1 summarizes these results for the solved instances. Figures 4.7 and 4.8 show the Pareto Front for each instance and method.

A point to highlight from Figs. 4.7 and 4.8, and Table 4.1, is that, for both instances, the Pareto front obtained by CPLEX is densely crowded. Additionally, it can be noted that the algorithms performed differently over the solved instances.

**Table 4.1** Quantity of non-dominated points for instances NOGC1 and NOGC2

| Instance name | $n$ | $k$ | CPLEX | NSGA | PSO |
|---|---|---|---|---|---|
| NOGC1 | 12 | 5 | 35 | 9 | 9 |
| NOGC2 | 15 | 8 | 10 | 14 | 6 |

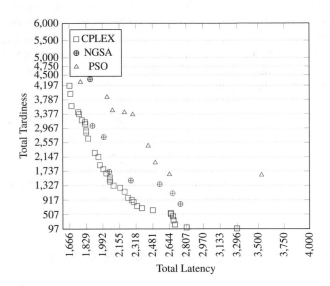

**Fig. 4.7** Pareto front, for instance, NOGC1

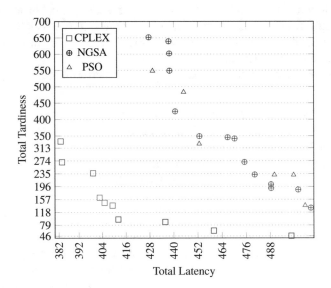

**Fig. 4.8** Pareto front, for instance, NOGC2

**Table 4.2** Elapsed CPU time (in seconds) for instances NOGC1 and NOGC2

| Instance name | Exact | NSGA | PSO |
|---|---|---|---|
| NOGC1 | 9774.481 | 0.177 | 0.238 |
| NOGC2 | 26,988.267 | 0.256 | 0.148 |

**Table 4.3** Maximum and average $k$-distances for the NOGC1 and NOGC2

| Instance name | Exact | | NSGA | | PSO | |
|---|---|---|---|---|---|---|
| | Max | Avg | Max | Avg | Max | Avg |
| NOGC1 | 0.321661 | 0.0464688 | 0.477647 | 0.204513 | 0.554544 | 0.177678 |
| NOGC2 | 0.505058 | 0.175595 | 0.206699 | 0.121096 | 0.412595 | 0.261747 |

**Table 4.4** Hypervolume for instances NOGC1 and NOGC2

| Instance name | Exact | NSGA | PSO |
|---|---|---|---|
| NOGC1 | 0.848594 | 0.690700 | 0.532383 |
| NOGC2 | 0.891554 | 0.317507 | 0.298918 |

In particular, for the instance NOGC1, the NSGA algorithm produced a front that is closer to the optimal front obtained by CPLEX. On the contrary, for instance NOGC2, both algorithms produced fronts that are far from the optimal front and, in particular, the NSGA produced a more dense front than the PSO.

Regarding the computational time, the summarized results are presented in Table 4.2. In this case, the exact method required almost 3 h for solving the NOGC1 instance (12 nodes), whereas, for the case of the instance NOGC2 (15 nodes), the time required to solve the instance was almost tripled. In particular, both metaheuristics required less than 1 s to solve the instances. In conjunction with the metric of the quantity of non-dominated points, this fact supports. the fact that the NSGA procedure performed better.

Regarding the density of the fronts, Table 4.3 shows the average $k$-distance value of all points on the efficient frontiers for each instance, with $k = 3$. As seen, the PSO provides a much better average of 3-distances than the NSGA. Although the exact method obtained the front with more density (minimum average $k$-distance), comparing the algorithms, the NSGA produced fronts with more density than the PSO. In particular, for the NOGC2 instance, it can be seen that the NSGA obtained the minimum values for the maximum and average distances.

To verify the efficiency of the NSGA, we used the hypervolume metric. Table 4.4 displays the obtained results. According to this metric, referring to the metaheuristic procedures, the NSGA algorithm provided greater hypervolume values than the PSO.

Finally, we used the coverage measure (considering only the strict domination). Tables 4.5 and 4.6 exhibit the results. In these tables, a value of $C(X', X'')$ equal to 1 means that all points in the estimated efficient frontier $X''$ are strictly dominated by points in the estimated efficient frontier $X'$. As expected, the exact method dominates both algorithms entirely in terms of the space covered. Regarding the metaheuristic procedures, for instance, NOGC1, the NSGA was able to generate

**Table 4.5** Coverage of two sets value, for instance, NOGC1

| X'/X'' | Exact | NSGA | PSO |
|---|---|---|---|
| Exact | 0 | 1 | 1 |
| NSGA | 0 | 0 | 0.888889 |
| PSO | 0 | 0.222222 | 0 |

**Table 4.6** Coverage of two sets value, for instance, NOGC2

| X'/X'' | Exact | NSGA | PSO |
|---|---|---|---|
| Exact | 0 | 1 | 1 |
| NSGA | 0 | 0 | 0.5 |
| PSO | 0 | 0.428571 | 0 |

points that dominate almost 88.89% of the ones generated by the PSO. For instance, NOGC2, the front of the NSGA dominates 50% of the points generated by the PSO.

We observe that for the instances NOCG1 and NOGC2 when the minimum value of latency is obtained, the maximum amount of total tardiness rises to 2.5 times the cost of latency, which can translate in a high level of customer dissatisfaction. On the contrary, when the minimum value of total tardiness is reached, the overall system's latency rises to 1.5 times the optimal (minimum) value. In this case, the increment of latency translates into a significant increase in energy consumption and, therefore, in the emission of greenhouse gas emissions. Also, the prioritization of the customers in the routes generates an unbalance in their demand, especially having routes where relatively few customers have significantly high amounts of demand than the rest.

Another aspect to highlight is that the decision-making process can be seen from two perspectives: On the one hand, the savings in tardiness costs can represent up to 67% of the total costs, whereas savings in latency produce savings up to 25% of the total costs. In other words, according to the objective function, if a preference must be defined *apriori*, tardiness must be more important than latency.

### 4.6.3 Experimental Results for Larger Instances

As mentioned above, since the formulation could not solve instances larger than 15 nodes, only the results obtained by the NSGA and PSO algorithms were compared. The complementary experimentation was conducted considering instances up to 100 nodes. According to the values reported for the multi-objective metrics, none of the algorithms outperformed each other. Tables 4.7, 4.8, 4.9, 4.10, 4.11, and 4.12 display the computational experimentation results.

In Table 4.7, column 1 displays the name of the instance. Columns 2 and 3 indicate the size in terms of the number of nodes and the number of routes. Columns 4 and 5 report the number of non-dominated solutions obtained by each algorithm. For Tables 4.9 and 4.10, column 1 shows the name of the instance, whereas columns 2 and 3 report the value of the corresponding algorithms over the evaluated metric.

**Table 4.7** Quantity of non-dominated points for the rest of the instances

| Instance name | $n$ | $k$ | NSGA | PSO |
|---|---|---|---|---|
| NOGC3 | 20 | 3 | 6 | 4 |
| NOGC4 | 22 | 8 | 8 | 6 |
| NOGC5 | 75 | 4 | 7 | 6 |
| NOGC6 | 75 | 7 | 5 | 6 |
| NOGC7 | 75 | 10 | 7 | 3 |
| NOGC8 | 75 | 7 | 11 | 7 |
| NOGC9 | 75 | 8 | 13 | 4 |
| NOGC10 | 100 | 5 | 8 | 6 |

**Table 4.8** Minimum and maximum values for both objectives functions for the remaining instances

| Instance name | Type of objective | NSGA | | PSO | |
|---|---|---|---|---|---|
| | | Min | Max | Min | Max |
| NOGC3 | Latency | 5327.35 | 7510.06 | 5568.42 | 7987.47 |
| | Tardiness | 4318.00 | 19,782.90 | 6446.31 | 16,733.70 |
| NOGC4 | Latency | 660.85 | 1102.98 | 800.06 | 978.52 |
| | Tardiness | 233.65 | 1159.62 | 14.87 | 137.33 |
| NOGC5 | Latency | 8275.60 | 10,376.60 | 8866.28 | 10,411.70 |
| | Tardiness | 49,416.40 | 66,396.30 | 36,109.80 | 27,515.30 |
| NOGC6 | Latency | 7366.98 | 9251.54 | 7491.19 | 8394.69 |
| | Tardiness | 20,883.80 | 31,319.70 | 18,804.10 | 32,214.4 |
| NOGC7 | Latency | 12,511.70 | 15,026.20 | 10,027.10 | 11,253.50 |
| | Tardiness | 82,739.9 | 12,4013.00 | 61,540.60 | 116,880.00 |
| NOGC8 | Latency | 13,683.50 | 16,203.90 | 12,560.60 | 14,359.10 |
| | Tardiness | 116,976.00 | 199,977.00 | 94,903.70 | 179,507.00 |
| NOGC9 | Latency | 9617.94 | 14,949.70 | 11,504.2 | 11,867.00 |
| | Tardiness | 69,126.70 | 130,754.00 | 80,684.60 | 150,941.00 |
| NOGC10 | Latency | 31,092.50 | 40,164.90 | 31,708.90 | 35,263.20 |
| | Tardiness | 343,722.00 | 530,071.00 | 341,296.00 | 541,012.00 |

Specifically, for Table 4.11, two columns are used to indicate the maximum and average $k$-distances for each procedure.

Following the same sequence for reporting the results, the first metric to be compared is the quantity of non-dominated points. Table 4.7 reports the size of the fronts obtained by each algorithm. According to the information there, the NSGA reported a higher quantity of non-dominated points. In some instances, the number of points reported almost doubled the amount of the ones obtained by the PSO.

In Table 4.8, the minimum and maximum values for each objective are shown. According to the information obtained, for most of the instances, the NSGA algorithm obtained better values for the total latency (except for the instances NOGC7 and NOGC8). On the other hand, the PSO produced better values of

**Table 4.9** Elapsed CPU time for the rest of the instances

| Instance name | NSGA | PSO |
|---|---|---|
| NOGC3 | 0.262 | 0.192 |
| NOGC4 | 0.768 | 0.163 |
| NOGC5 | 6.415 | 0.193 |
| NOGC6 | 9.919 | 0.195 |
| NOGC7 | 62.262 | 0.251 |
| NOGC8 | 47.039 | 0.259 |
| NOGC9 | 48.816 | 0.260 |
| NOGC10 | 96.541 | 0.292 |

**Table 4.10** Hypervolume for the rest of the instances

| Instance name | NSGA | PSO |
|---|---|---|
| NOGC3 | 0.857117 | 0.439723 |
| NOGC4 | 0.412973 | 0.672329 |
| NOGC5 | 0.727375 | 0.443356 |
| NOGC6 | 0.562751 | 0.547212 |
| NOGC7 | 0.182504 | 0.893468 |
| NOGC8 | 0.375955 | 0.875043 |
| NOGC9 | 0.750632 | 0.494252 |
| NOGC10 | 0.81878 | 0.639694 |

**Table 4.11** Maximum and average $k$-distances for the rest of the instances

| Instance name | NSGA | | PSO | |
|---|---|---|---|---|
| | Max | Avg | Max | Avg |
| NOGC3 | 0.957653 | 0.411645 | 0.841626 | 0.220280 |
| NOGC4 | 0.563797 | 0.302869 | 0.243252 | 0.121841 |
| NOGC5 | 0.773372 | 0.317767 | 0.401314 | 0.250991 |
| NOGC6 | 0.785308 | 0.572339 | 0.690433 | 0.362093 |
| NOGC7 | 0.390827 | 0.197992 | 0.919165 | 0.775966 |
| NOGC8 | 0.348243 | 0.197771 | 0.600042 | 0.246500 |
| NOGC9 | 0.651263 | 0.160927 | 0.770504 | 0.281498 |
| NOGC10 | 0.77901 | 0.314597 | 0.486934 | 0.305644 |

tardiness for most of the instances. The vehicle routes that provide minimum economic cost are not necessarily environment-friendly routes or vice versa.

Regarding the performance of the algorithms, it can be noticed that both algorithms showed a mixed behavior. To better illustrate this, the fronts of the instances NOGC8 and NOG10 are displayed in Figs. 4.9 and 4.10.

Due to this, the metric of the execution time was evaluated to verify if any of the procedures performs faster. Table 4.9 displays the elapsed CPU time for the best execution.

As seen, the NSGA increased the required time substantially as the size of the instances increased. On the contrary, the PSO consistently outperforms NSGA in terms of the execution time.

**Table 4.12** Coverage of two sets value for the rest of the instances

| Instance name | X′/X″ | Exact | |
| --- | --- | --- | --- |
| | | NSGA | PSO |
| NOGC3 | NSGA | 0 | 1 |
| | PSO | 0 | 0 |
| NOGC4 | NSGA | 0 | 0 |
| | PSO | 0.625 | 0 |
| NOGC5 | NSGA | 0 | 1 |
| | PSO | 0 | 0 |
| NOGC6 | NSGA | 0 | 0.833333 |
| | PSO | 0.4 | 0 |
| NOGC7 | NSGA | 0 | 0 |
| | PSO | 1 | 0 |
| NOGC8 | NSGA | 0 | 0 |
| | PSO | 1 | 0 |
| NOGC9 | NSGA | 0 | 1 |
| | PSO | 0 | 0 |
| NOGC10 | NSGA | 0 | 0.833333 |
| | PSO | 0.125 | 0 |

The third metric regards the hypervolume. The results of the algorithm are displayed in Table 4.10. There, it cannot be confirmed that any algorithm outperforms each other. What can be confirmed is that NSGA produced higher values of hypervolume for 5 out of 8 instances. However, for the instances where the PSO obtained better values, the difference against the NSGA was larger.

Regarding the density of the frontiers, Table 4.11 shows the results obtained. From these results, it can be noticed that, the PSO produced lower values for the maximum distances (more compactness). However, the NSGA algorithm produced better values for the average distances.

Finally, Table 4.12 reports the values obtained for the set coverage metric. The first column refers to the name of the instance, and the rest show comparisons in coverage between the algorithms. Again, the NSGA performed better than the PSO. However, for instances in which PSO was better than the others, the procedure generated a front that entirely covered the one obtained by the NSGA. It confirms that none of the algorithms dominates.

In summary, we conclude that both metaheuristics provided good results in a reasonable computational time, but none of them consistently outperforms each other.

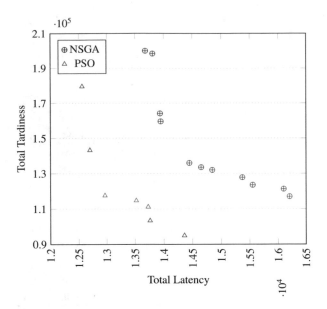

**Fig. 4.9** NSGA and PSO Pareto fronts, for instance, NOGC8

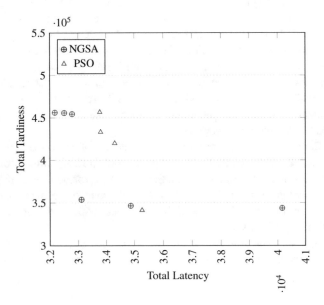

**Fig. 4.10** NSGA and PSO Pareto fronts, for instance, NOGC10

## 4.7   Conclusions and Future Work

This chapter addressed the bi-objective Cumulative Capacitated Vehicle Routing Problem. The problem arises in a commercial context, like delivering perishable goods when these are differentiated based on priority indexes. In the case of a pooled transportation service might help to estimate the trade-off between delivering the orders in the same sequence as the customers board the vehicle and the minimum arrival time of the system. For this problem, a mixed-integer programming formulation and two metaheuristic algorithms were developed. A commercial optimization software solved the model for small size instances. In contrast, the algorithms showed their effectiveness by providing feasible results in a reasonable amount of computational time.

The studies described in the literature review imply that the $CO_2$ emission and energy consumption vary with the traveling distance and gross vehicle weight. Optimizing the routing in terms of distance leads to an environmentally efficient operation. Additionally, the formulation presented in this chapter has the advantage of dealing with a heterogeneous fleet (vehicles). Since the fuel consumption pattern is closely related to the vehicle's condition the flexibility offered by using different types of vehicles may result in a higher reduction of fuel consumption. Another benefit of the proposed approach is that the managers can determine the appropriate priority indexes in order to accomplish with the environmental regulations. Regarding social concerns, in many cases, the customers located far from the depot are considered less attractive for delivery, producing discrimination that can be counteracted by assigning higher priority indexes to those customers. Also, the appropriate assignment of priority indexes can be used for waste collection systems, to prioritize sites with greater demand or hazardous waste. On the other hand, tardiness is critical for refrigerated freights, where a longer time in the route implies a higher $CO_2$ emission and energy consumption. These situations can be addressed by the model proposed with the appropriate values and balance of parameters.

Regarding the algorithms, these showed their efficiency by providing good quality fronts for the small-sized instances. Also, for larger instances, both algorithms provided good values for the multi-objective metrics evaluated. Although none of the algorithms outperformed each other, the NSGA obtained fronts with a higher quantity of points, more density, and more coverage of the sets. However, PSO showed a better performance in terms of computational time.

In summary, all procedures provided a positive contribution to a more sustainable balance between economic, environmental, and social objectives. Our results are useful to the business in terms of care of the environment and allow the firm to gain a valuable competitive advantage by showing green credentials.

Future research lines include the design of routes using congested environments with travel speeds, time windows as priority metrics, and capacity constraints. In addition, other factors to consider can be labor costs or balance of the total traveled distance among routes, which seem to dominate the overall cost.

**Acknowledgments** This work was supported by the Universidad Panamericana through the grant "Fondo Fomento a la Investigación UP 2019", under project code UP-CI-2019-ING-GDL-08.

# References

1. Srivastava, S.K.: Int. J. Manag. Rev. **9**, 53–80 (2007). https://onlinelibrary.wiley.com/doi/abs/10.1111/j.1468-2370.2007.00202.x
2. Engin, B.E., Martens, M., Paksoy, T.: Lean and Green Supply Chain Management: A Comprehensive Review, pp. 1–38. Springer International Publishing, Cham (2019). https://doi.org/10.1007/978-3-319-97511-5_1
3. Ngueveu, S.U., Prins, C., Calvo, R.W.: Comput. Oper. Res. **37**, 1877–1885 (2010)
4. Martínez-Salazar, I., Angel-Bello, F., Alvarez, A.: J. Oper. Res. Society **66**, 1312–1323 (2015)
5. Rivera, J.C., Afsar, H.M., Prins, C.: Comput. Optim. Appl. **61**, 159–187 (2015)
6. Gaur, D.R., Mudgal, A., Singh, R.R.: Improved approximation algorithms for cumulative VRP with stochastic demands. Discret. Appl. Math. **280**, 133–143 (2020). https://doi.org/10.1016/j.dam.2018.01.012
7. Lalla-Ruiz, E., Voß, S.: Optim. Lett. 1–21 (2019)
8. Kara, İ., Kara, B.Y., Kadri Yetiş, M.: Vehicle Routing Problem. IntechOpen, London (2008)
9. Rivera, J.C., Afsar, H.M., Prins, C.: Eur. J. Oper. Res. **249**, 93–104 (2016)
10. Nucamendi-Guillén, S., Angel-Bello, F., Martínez-Salazar, I., Cordero-Franco, A.E.: Expert Syst. Appl. **113**, 315–327 (2018)
11. Lysgaard, J., Wøhlk, S.: Eur. J. Oper. Res. **236**, 800–810 (2014)
12. Ribeiro, G.M., Laporte, G.: Comput. Oper. Res. **39**, 728–735 (2012)
13. Ozsoydan, F.B., Sipahioglu, A.: Optimization **62**, 1321–1340 (2013)
14. Ke, L., Feng, Z.: Comput. Oper. Res. **40**, 633–638 (2013)
15. Lin, C., Choy, K., Ho, G., Chung, S., Lam, H.: Expert Syst. Appl. **41**, 1118–1138 (2014). http://www.sciencedirect.com/science/article/pii/S095741741300609X
16. Karagul, K., Sahin, Y., Aydemir, E., Oral, A.: A Simulated Annealing Algorithm Based Solution Method for a Green Vehicle Routing Problem with Fuel Consumption, pp. 161–187. Springer International Publishing, Cham (2019). https://doi.org/10.1007/978-3-319-97511-5_6
17. Kara, İ., Kara, B.Y., Yetis, M.K. In: Dress, A., Xu, Y., Zhu, B. (eds.) Combinatorial Optimization and Applications, pp. 62–71. Springer, Berlin (2007)
18. Palmer, A.: School of Management. Cranfield University, Cranfield (2007)
19. Sbihi, A., Eglese, R.W.: 4OR **5**, 99–116 (2007). https://doi.org/10.1007/s10288-007-0047-3
20. Kuo, Y.: Comput. Ind. Eng. **59**, 157–165 (2010). http://www.sciencedirect.com/science/article/pii/S0360835210000835
21. Maden, W., Eglese, R., Black, D.: J. Oper. Res. Society **61**, 515–522 (2010). https://doi.org/10.1057/jors.2009.116
22. Figliozzi, M.: Transport. Res. Record **2197**, 1–7 (2010). https://doi.org/10.3141/2197-01
23. Urquhart, N., Scott, C., Hart, E. In: Di Chio, C., Brabazon, A., Di Caro, G.A., Ebner, M., Farooq, M., Fink, A., Grahl, J., Greenfield, G., Machado, P., O'Neill, M., Tarantino, E., Urquhart, N. (eds.) Applications of Evolutionary Computation, pp. 421–430. Springer, Berlin (2010)
24. Bektaş, T., Laporte, G.: Transp. Res. B Methodol. **45**, 1232–1250 (2011). Supply chain disruption and risk management. http://www.sciencedirect.com/science/article/pii/S019126151100018X
25. Faulin, J., Juan, A., Lera, F., Grasman, S.: Procedia - Social and Behavioral Sciences **20**, 323–334 (2011). The State of the Art in the European Quantitative Oriented Transportation and Logistics Research — 14th Euro Working Group on Transportation & 26th Mini Euro Conference & 1st European Scientific Conference on Air Transport. http://www.sciencedirect.com/science/article/pii/S1877042811014182

26. Ubeda, S., Arcelus, F., Faulin, J.: Int. J. Prod. Eco. **131**, 44–51 (2011). Innsbruck 2008. http://www.sciencedirect.com/science/article/pii/S092552731000174X
27. Suzuki, Y.: Transp. Res. Part D: Transp. Environ. **16**, 73–77 (2011). http://www.sciencedirect.com/science/article/pii/S1361920910001239
28. Demir, E., Bektaş, T., Laporte, G.: Eur. J. Oper. Res. **223**, 346–359 (2012). http://www.sciencedirect.com/science/article/pii/S0377221712004997
29. Jemai, J., Zekri, M., Mellouli, K. In: Hao, J.-K., Middendorf, M. (eds.) Evolutionary Computation in Combinatorial Optimization, pp. 37–48. Springer, Berlin (2012)
30. Erdoğan, S., Miller-Hooks, E.: Transport Res E-Log **48**, 100–114 (2012). Select Papers from the 19th International Symposium on Transportation and Traffic Theory. http://www.sciencedirect.com/science/article/pii/S1366554511001062
31. Xiao, Y., Zhao, Q., Kaku, I., Xu, Y.: Comput. Oper. Res. **39**, 1419–1431 (2012). http://www.sciencedirect.com/science/article/pii/S0305054811002450
32. Huang, Y., Shi, C., Zhao, L., Woensel, T.V. (eds.) Proceedings of 2012 IEEE International Conference on Service Operations and Logistics, and Informatics, pp. 302–307. https://doi.org/10.1109/SOLI.2012.6273551
33. Ramos, T.R.P., Gomes, M.I., Barbosa-Póvoa, A.P.: Minimizing CO2 emissions in a recyclable waste collection system with multiple depots. In: 2012 EUROMA/POMS joint conference pp. 1-5
34. Omidvar, A., Tavakkoli-Moghaddam, R.: Sustainable vehicle routing: strategies for congestion management and refueling scheduling, pp. 1089–1094. IEEE, Piscataway (2012). https://doi.org/10.1109/EnergyCon.2012.6347732
35. Li, J.: J. Comput. **7**, 3020–3027 (2012). https://doi.org/10.4304/jcp.7.12.3020-3027
36. Jabali, O., Van Woensel, T., de Kok, T.: Prod. Oper. Manag. **21**, 1060–1074 (2012). https://doi.org/10.1111/j.1937-5956.2012.01338.x
37. Franceschetti, A., Honhon, D., Woensel, T.V., Bektaş, T., Laporte, G.: Transp. Res. B Methodol. **56**, 265–293 (2013). http://www.sciencedirect.com/science/article/pii/S0191261513001446
38. Peiying, Y., Jiafu, T., Yu, Y.: Based on Low Carbon Emissions Cost Model and Algorithm for Vehicle Routing and Scheduling in Picking Up and Delivering Customers to Airport Service, pp. 1693–1697. IEEE, Piscataway (2013). https://doi.org/10.1109/CCDC.2013.6561203
39. Kwon, Y.-J., Choi, Y.-J., Lee, D.-H.: Transp. Res. Part D: Transp. Environ. **23**, 81–89 (2013). http://www.sciencedirect.com/science/article/pii/S1361920913000643
40. Yasin, M., Yu, V.F. In: Lin, Y.-K., Tsao, Y.-C., Lin, S.-W. (eds.) Proceedings of the Institute of Industrial Engineers Asian Conference 2013, pp. 1261–1269. Springer, Singapore (2013)
41. Küçükoğlu, I., Ene, S., Aksoy, A., Öztürk, N.: Int. J. Comput. Eng. Res. **3**, 16–23 (2013)
42. Pradenas, L., Oportus, B., Parada, V.: Expert Syst. Appl. **40**, 2985–2991 (2013). http://www.sciencedirect.com/science/article/pii/S0957417412012559
43. Kopfer, H.W., Schönberger, J., Kopfer, H.: Flex. Serv. Manuf. J. **26**, 221–248 (2014). https://doi.org/10.1007/s10696-013-9180-9
44. Treitl, S., Nolz, P.C., Jammernegg, W.: Flex. Serv. Manuf. J. **26**, 143–169 (2014). https://doi.org/10.1007/s10696-012-9158-z
45. Úbeda, S., Faulin, J., Serrano, A., Arcelus, F.J.: Lecture Notes Manag. Sci. **6**, 141–149 (2014)
46. Taha, M., Fors, N., Shoukry, A. (2014)
47. Ayadi, R., ElIdrissi, A.E., Benadada, Y., El Hilali Alaoui, A. In: 2014 International Conference on Logistics Operations Management, pp. 148–154. https://doi.org/10.1109/GOL.2014.6887432
48. Adiba, E.E., Aahmed, E.A., Youssef, B.: In: 2014 International Conference on Logistics Operations Management, pp. 161–167. https://doi.org/10.1109/GOL.2014.6887434
49. Montoya, A., Guéret, C., Mendoza, J.E., Villegas, J.G.: Transp. Res. C: Emerg. Technol. **70**, 113–128 (2016). http://www.sciencedirect.com/science/article/pii/S0968090X15003320
50. Ene, S., Küçükoğlu;, I., Aksoy, A., Öztürk, N.: Int. J. Veh. Desig. **71**, 75–102 (2016). https://www.inderscienceonline.com/doi/abs/10.1504/IJVD.2016.078771

51. ÇağrıKoç, Karaoglan, I.: Appl. Soft Comput. **39**, 154–164 (2016). http://www.sciencedirect. com/science/article/pii/S1568494615007085
52. Afshar-Bakeshloo, M., Mehrabi, A., Safari, H., Maleki, M., Jolai, F.: J. Ind. Eng. Int. **12**, 529–544 (2016). https://doi.org/10.1007/s40092-016-0163-9
53. Arango Gonzalez, D.S., Olivares-Benitez, E., Miranda, P.A.: Adv. Oper. Res. **2017**, 11 (2017). https://doi.org/10.1016/10.1155/2017/4093689
54. Andelmin, J., Bartolini, E.: Trans. Sci. **51**, 1288–1303 (2017) . https://doi.org/10.1287/trsc. 2016.0734
55. Andelmin, J., Bartolini, E.: Comput. Oper. Res. **109**, 43–63 (2019). http://www.sciencedirect. com/science/article/pii/S0305054819301017
56. Yu, V.F., Redi, A.P., Hidayat, Y.A., Wibowo, O.J.: Appl. Soft Comput. **53**, 119–132 (2017) . http://www.sciencedirect.com/science/article/pii/S1568494616306524
57. Cooray, P.L.N.U., Rupasinghe, T.: J. Ind. Eng. **2017**, 1–13 (2017). https://doi.org/10.1155/ 2017/3019523
58. Sawik, B., Faulin, J., Perez-Bernabeu, E.: Transport. Res. Proc. **22**, 305–313 (2017). https:// doi.org/10.1016/j.trpro.2017.03.037
59. Toro, E.M., Franco, J.F., Echeverri, M.G., aes, F.G.G.: Comput. Ind. Eng. **110**, 114–125 (2017). http://www.sciencedirect.com/science/article/pii/S0360835217302176
60. Toro, E., Franco, J., Granada-Echeverri, M., Guimarães, F., Gallego Rendón, R.A.: Int. J. Ind. Eng. Comput. **8**, 203–216 (2016). https://doi.org/10.5267/j.ijiec.2016.10.001
61. Hamid Mirmohammadi, S., Babaee Tirkolaee, E., Goli, A., Dehnavi-Arani, S.: Iran Univ. Sci. Technol. **7**, 143–156 (2016)
62. de Oliveira da Costa, P.R., Mauceri, S., Carroll, P., Pallonetto, F.: Electron Notes Discrete Math. **64**, 65–74 (2018). 8th International Network Optimization Conference – INOC 2017. http://www.sciencedirect.com/science/article/pii/S1571065318300088
63. Tirkolaee, E.B., Hosseinabadi, A.A.R., Soltani, M., Sangaiah, A.K., Wang, J.: Sustainability **10**, 1–21 (2018). https://ideas.repec.org/a/gam/jsusta/v10y2018i5p1366-d143612.html
64. Macrina, G., Pugliese, L.D.P., Guerriero, F., Laporte, G.: Comput. Oper. Res. **101**, 183–199 (2019). http://www.sciencedirect.com/science/article/pii/S0305054818301965
65. Wang, L., Lu, J.: IEEE/CAA J. Automat. Sin. **6**, 516–526 (2019). https://doi.org/10.1109/JAS. 2019.1911405
66. Li, Y., Soleimani, H., Zohal, M.: J. Clean. Prod. **227**, 1161–1172 (2019). http://www. sciencedirect.com/science/article/pii/S0959652619308790
67. Yu, Y., Wang, S., Wang, J., Huang, M.: Transport. Res. B: Methodol. **122**, 511–527 (2019). http://www.sciencedirect.com/science/article/pii/S0191261518308944
68. Dukkanci, O., Kara, B.Y., Bektaş, T.: Comput. Oper. Res. **105**, 187–202 (2019). http://www. sciencedirect.com/science/article/pii/S0305054819300218
69. Angel-Bello, F., Alvarez, A., García, I.: Appl. Math. Model. **37**, 2257–2266 (2013). http:// www.sciencedirect.com/science/article/pii/S0307904X12003459
70. Mavrotas, G.: Appl. Math. Comput. **213**, 455–465 (2009) . http://www.sciencedirect.com/ science/article/pii/S0096300309002574
71. Holland, J.H. et al.: Adaptation in Natural and Artificial Systems: An Introductory Analysis with Applications to Biology, Control, and Artificial Intelligence. MIT Press, Cambridge (1992)
72. Prins, C.: Comput. Oper. Res. **31**, 1985–2002 (2004)
73. Srinivas, N., Deb, K.: Evol. Comput. **2**, 221–248 (1994). https://doi.org/10.1162/evco.1994.2. 3.221
74. Deb, K., Pratap, A., Agarwal, S., Meyarivan, T.: IEEE Trans. Evol. Comput. **6**, 182–197 (2002)
75. Martínez-Salazar, I.A., Molina, J., Ángel-Bello, F., Gómez, T., Caballero, R.: Eur. J. Oper. Res. **234**, 25–36 (2014)
76. Kennedy, J., Eberhart, R. In: Proceedings of ICNN'95 – International Conference on Neural Networks, vol. 4, pp. 1942–1948. https://doi.org/10.1109/ICNN.1995.488968
77. Chen, R.-M., Shen, Y.-M., Hong, W.-Z.: Exp. Syst. Appl. **138**, 112833 (2019). http://www. sciencedirect.com/science/article/pii/S0957417419305354

78. Li, X., Clerc, M.: Swarm Intelligence, pp. 353–384. Springer International Publishing, Cham (2019). https://doi.org/10.1007/978-3-319-91086-4_11
79. Talbi, E.-G.: Metaheuristics: From Design to Implementation, vol. 74. John Wiley & Sons, Hoboken (2009)
80. Okulewicz, M., Mańdziuk, J.: Swarm Evol. Comput. **48**, 44–61 (2019). http://www.sciencedirect.com/science/article/pii/S2210650218306114
81. Koulaeian, M., Seidgar, H., Kiani, M., Fazlollahtabar, H.: Int. J. Ind. Eng. Theory Appl. Pract. **22**, 223–242 (2015). http://journals.sfu.ca/ijietap/index.php/ijie/article/view/1379
82. Chunyu, R., Xiaobo, W. (eds.) 2010 International Conference on Artificial Intelligence and Computational Intelligence, vol. 1, pp. 552–555. https://doi.org/10.1109/AICI.2010.121
83. Gillett, B.E., Johnson, J.G.: Omega **4**, 711–718 (1976)
84. Augerat, P., Belenguer, J.M., Benavent, E., Corberán, A., Naddef, D., Rinaldi, G.: Computational results with a branch and cut code for the capacitated vehicle routing problem, Technical Report, IMAG (1995)
85. Zitzler, E., Laumanns, M., Thiele, L., EUROGEN 2001: Evolutionary Methods for Design, Optimization and Control with Applications to Industrial Problems, pp. 95–100 (2000)
86. Zitzler, E., Thiele, L.: IEEE Trans. Evol. Comput. **3**, 257–271 (1999)

# Chapter 5
# Solution of a Real-Life Vehicle Routing Problem with Meal Breaks and Shifts

**Çiğdem Karademir, Ümit Bilge, Necati Aras, Gökay Burak Akkuş, Göksu Öznergiz, and Onur Doğan**

**Abstract** Last mile delivery remains to be a focal point in the operations of logistics service providers. As more customers are willing to pay a premium price for receiving their parcel within a specific time window the Vehicle Routing Problem with Time Windows (VRPTW) has been an active research area. However, realistic problem features such as the existence of multiple working shifts and the meal breaks of the delivery personnel assigned to each shift have not received sufficient attention in the literature. The existence of these additional requirements gives rise to the VRPTW including Breaks and Shifts (VRPTW-BS) which is the main focus of the present study. The fact that the time windows of some customers overlap with several shifts brings additional complexity since the problem cannot be decomposed with respect to shifts. We develop three methods for the solution of the VRPTW-BS. The first method involves the formulation of a mixed-integer linear programming model, which can be solved to optimality for relatively small instances with up to 25 customers and with small optimality gaps up to 35 customers. The main use of this model is to assess the quality of the solutions obtained by the proposed two-phase heuristic that is based on generating an initial solution and improving it using local search methods. For larger instances with more than 1000 customers, we also devise another heuristic based on partitioning the customers into a number of clusters and then applying our heuristic for each cluster. The results obtained on instances of different size indicate that the proposed heuristic yields solutions of good quality within acceptable computation times.

Ç. Karademir (✉) · Ü. Bilge · N. Aras
Dept. of Industrial Engineering, Boğaziçi University, İstanbul, Turkey
e-mail: cigdem.karademir@boun.edu.tr; bilge@boun.edu.tr; arasn@boun.edu.tr

G. B. Akkuş · G. Öznergiz · O. Doğan
Ekol Logistics, İstanbul, Turkey
e-mail: gokay.akkus@ekol.com; goksu.oznergiz@ekol.com; onur.dogan@ekol.com

© Springer Nature Switzerland AG 2020
H. Derbel et al. (eds.), *Green Transportation and New Advances in Vehicle Routing Problems*, https://doi.org/10.1007/978-3-030-45312-1_5

131

## 5.1 Introduction

Last mile delivery has a central importance in the operations of logistic service providers (LSPs) worldwide. This importance will continue to increase in the future due to the boost in the number of people who make shopping online. It is estimated that e-commerce sales worldwide reach 3.45 trillion US dollars in 2019 and the same figure is expected to rise to 4.88 trillion US dollars in 2021.[1] 1.92 billion people will buy at least one item online in 2019.[2] As pointed out in the Digital Turkey 2019 E-commerce report,[3] the number of people who make purchases on the Internet was 39.3 million in 2018, and the amount of money spent became 11.5 billion US dollars. This corresponds to 44% increase from previous year in terms of local currency. As a result, it appears that a growing trend will be observed in the number of home deliveries in the following years all over the world.

This study is motivated by the request of an LSP operating in Turkey. This company offers many logistics services, and the line of business giving rise to the present study is the last mile delivery operation that aims to ensure timely delivery of products ordered by the customers at major e-tailers. An emerging trend in this sector is the customers' behavior in receiving the deliveries: more customers are willing to pay a premium price for receiving their parcel within a specific time window because they do not want the parcel to arrive when they are not at home. Besides this tendency in customer behavior, there are also some operational requirements to be considered by the LSP, which further complicate planning the activities of the last mile delivery. First, the delivery personnel of the company work in two shifts during the day. The first shift starts at 8:15 am and finishes at 3:45 pm, while the working hours of the second shift span the time between 2:45 pm and 10:15 pm. This means that the two shifts have overlapping service hours. The second requirement on the operational side is the existence of meal breaks in the shifts. The duration of the breaks is 30 min, and they take place between 12:00 pm and 12:30 pm in the first shift, and between 6:30 pm and 7:00 pm in the second shift. It is worth noticing that specific locations are not designated for meal breaks, and they can take place between two customer locations.

The goal of the LSP is to carry out the delivery operation with the smallest fleet of vehicles. This means that the company is less concerned with the total distance traveled by the vehicles. Most probably, this strategy is due to the fact that the fixed cost of a vehicle is much more significant than the reduction in the traveling distance or fuel cost that can be gained by increasing the fleet size by an additional vehicle. Hence, the problem addressed by the LSP is a variant of the Vehicle Routing Problem with Time Windows (VRPTW) which includes meal breaks and shifts. We refer to this problem as VRPTW-BS.

---

[1] https://www.statista.com/statistics/379046/worldwide-retail-e-commerce-sales/ accessed on 26 Aug 2019.

[2] https://hostingfacts.com/internet-facts-stats/ accessed on 26 Aug 2019.

[3] https://www.fundalina.com/dijital-turkiye-2019-e-ticaret-raporu-aciklandi/ accessed on 26 Aug 2019.

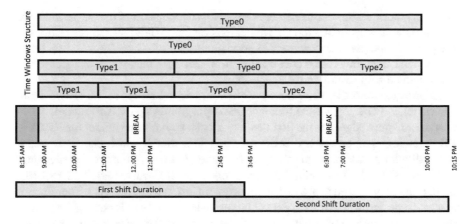

**Fig. 5.1** The structure of time windows and shifts

As a matter of fact, the LSP defines several time windows for its customers and some of these time windows indeed overlap with both shifts. There are also time windows that entirely remain within the interval of one shift. The largest time window covers the daily working hours of the LSP. Figure 5.1 illustrates the nature of the time windows as well as the shifts. In this figure, Type 1 and Type 2 refer to the time windows remaining in a shift's interval, whereas Type 0 time windows overlap with both shifts. It is worth mentioning that the existence of Type 0 time windows brings additional complexity to the solution of VRPTW-BS since the problem cannot be decomposed with respect to shifts. Therefore, in addition to the classical routing and service-start decisions that need to be made in VRPTW, decisions related to customer-to-shift assignment decisions should also be taken into account. Moreover, the solutions must have the property that the waiting time of a vehicle before it starts service at a customer does not exceed a certain threshold value. Notice that the waiting of a vehicle occurs when it arrives at a customer before the earliest start time of the service, that is the beginning of the time window.

Other characteristics of the problem dealt with in this study can be summarized as follows. There is a single depot, and the number of customers that need to be visited daily is above 1000. Moreover, the distances between the customers as well as between the depot and customers are asymmetric. Another important point worth mentioning is that on the basis of the information provided by the logistics service provider with which this study is carried out it is possible to assume that the fleet consists of homogeneous vehicles and that the vehicles are uncapacitated. The main reason of the latter assumption is the existence of time windows for customers, the duration of the shifts, meal breaks, and the non-zero service times at customers restrict the maximum number of customers that can be served by a vehicle in a shift. Moreover, the vehicles owned by the LSP have sufficient capacity to easily carry and deliver all the items for customers whose number is limited due to aforementioned reasons.

We develop a heuristic method for the solution of the VRPTW-BS, which turns out to be a rich VRP. This heuristic consists of two stages. In the first stage, a construction heuristic is utilized that is based on an extension of the classical Clarke and Wright savings heuristic [1]. This extension can handle all the restrictions of the problem and generate solutions of good quality within short computation times. The second stage involves a post-optimization step that makes use of local search implemented with several move operators.

In order to assess the performance of our heuristic referred to as TPH (Two-Phase Heuristic), we develop a mixed-integer linear programming (MILP) model. To the best of our knowledge, it is the first mathematical programming model for VRPTW-BS including meal breaks and shifts together. This model which can be solved to optimality for problems up to 25 nodes. However, the size of the problem instances to be solved by the LSP on a daily basis exceeds 1000 customers, and an MILP model cannot be solved to optimality for problem instances of this size. Therefore, to assess the performance of the TPH for larger instances we opt for another approach that has the capability of solving larger-sized instances. The idea of this approach is based on a well-known technique, decomposition, which is frequently implemented in the operations research literature for large problem instances. We decompose a given VRPTW-BS instance into smaller instances by means of $k$-means clustering, and apply our heuristic for the customer in each cluster.

The main contributions of this study can be summarized as follows:

- VRPTW-BS, which includes multiple working shifts and meal breaks simultaneously, is defined for the first time in the literature.
- A mathematical programming formulation is developed for the VRPTW-BS and solved for instances up to 35 customers with optimality gaps less than 7%.
- A two-phase heuristic called TPH is proposed which can solve instances with more than 1000 customers.

The remainder of the chapter is organized as follows. Section 5.2 contains the literature review. Section 5.3 presents the mathematical models involved. Section 5.4 includes the proposed heuristic TPH, while Sect. 5.5 consists of the computational results. Section 5.6 finishes the paper with concluding remarks and studies that are planned.

## 5.2 Literature Review

As mentioned in Coelho et al. [2], there are relatively few studies in the literature that address VRPTW with lunch breaks (VRPTWLB) although considerable research effort has been put for the VRPTW [3]. A Waste Collection Vehicle Routing Problem with Time Windows is considered in Buhrkal et al. [4], where the goal is to find routes for waste collection trucks so as to minimize the cost of the routes. There exist two differences between this problem and the classical VRPTW. First, the trucks must also visit disposal sites within the routes whenever they are full to unload the waste they collected. It is possible that a vehicle visits the same disposal sites more than once. Second, drivers give two types of breaks: rest breaks and lunch breaks. After 4.5 h of driving a rest break of 45 min must be given and a lunch break of minimum 30 min must be held once during the day. The authors propose an adaptive large neighborhood search heuristic to solve the problem. Although they also present an MILP, they do not solve it. Sahoo et al. [5] focus on a similar waste collection problem that includes a single node for lunch break. An MILP is proposed for a simplified version of their problem, but no result is presented with respect to the solution of this mathematical model. An iterative two-phase algorithm is developed where an initial solution is generated in the first step which is then improved in the second step by an extended insertion algorithm as well as a simulated-annealing metaheuristic that makes use of the cross exchange local search method proposed by Taillard et al. [6]. Another study which uses a single node representation for meal break is due to Sze [7]. The authors solve a real-life in-flight food loading operation as a multi-trip VRPTWLB in which food items are loaded to aircrafts within their transit time. The loading personnel should have their meal breaks in lounge areas designated as a node that has a time window as large as possible to give a break. A mathematical model is formulated for the problem and a two-stage heuristic is proposed. The heuristic solutions are better than those of the company, but relatively worse than the optimal solutions.

The same problem is investigated by Kim et al. [8] and solved by means of a clustering-based route building algorithm that is implemented with and without simulated annealing. Benjamin and Beasley [9] also study the same waste collection problem as Kim et al. [8]. They present three different metaheuristics based on variable neighborhood search, tabu search, and variable neighborhood search with tabu search. These metaheuristics provide solutions of similar quality, but the average distance of these solutions is smaller and the number of vehicles is higher than those of Kim et al. [8]. Therefore, to reduce the number of vehicles they introduce a vehicle reduction procedure which gives rise to the same number of vehicles. The VRPTWLB is addressed in Coelho et al. [2], where both a new mathematical formulation and multi-start randomized local search heuristic are introduced for it. The developed MILP can only provide optimal solutions for very small instances with five and 10 customers. An instance with 20 customers can be solved with 52% optimality gap in 1 h computation time. The reason lies in the fact that the MILP formulation does not use binary variables to indicate the customer nodes after which a lunch break is given, rather the number of customer nodes is doubled in order to represent the break nodes, which significantly increases the number of routing variables. An important result of the paper is the demonstration of integrating drivers' breaks into the model. The authors show that solving the VRPTWLB without explicitly taking into account the breaks and insert them into the obtained routes in a post-processing step may generate infeasible solutions. A unified hybrid genetic search metaheuristic is devised by Vidal et al. [10] to solve the VRPTWLB and other rich VRP problems. Lunch breaks can be given in a route at a location that can be selected from among a set of potential sites or the location of the break can be unrestricted. The building blocks of this heuristic that are problem-independent are unified local search, genetic operators, and diversity management methods, while there also exist a limited number of problem-specific components.

Within the context of waste collection including multiple depots and intermediate facilities, a mathematical model is formulated by Markov et al. [11] in which breaks are defined on arcs and the number of vehicles is fixed. Exact solutions are obtained by solving this model up to 15 customers while multiple neighborhood search instances with up to 288 customers and six intermediate facilities. Furthermore, there are several recent studies on electrical vehicle routing with time windows (EVRPTW) [12]. However lunch break and multiple working shift features have not been incorporated yet in this line of research.

## 5.3 Mathematical Model

As pointed out earlier, the company has two objectives which are hierarchical in nature. The objective function which has priority over the other one is minimizing the number of vehicles. The second objective function is related to the minimization of the total traveling distance. Since the problem is a rich VRP with many constraints including time windows, breaks, and working shifts, it is not clear how many vehicles are required to obtain a feasible solution for the VRPTW-BS. Note that, distance minimization does not guarantee minimum number of vehicles in general, and for a given problem, the total distance with a larger number of routes might be smaller than the total distance with a smaller number of routes. Therefore, we first formulate an MILP model in order to determine the minimum number of vehicles required. This model is referred to as MBS1. After having obtained the minimum number of vehicles from the optimal solution of MBS1, we utilize another MILP model called MBS2 which has the objective of minimizing the total distance traveled by the vehicles. In other words, the second model takes the number of vehicles necessary for the operation as an input parameter. In formulating the two models, we use the following index sets, parameters, and decision variables.

We let $N$ denote the set of customer nodes at which a delivery has to be made. All vehicles depart from a depot indexed by node 0 and the set $N_0 = N \cup \{0\}$ represents all the nodes included in the model. The last index set $\mathcal{K} = \{1, 2\}$ is related to the set of shifts. The delivery personnel of the company work in two shifts. The first shift starts at 8:15am and finishes at 3:45pm, while the working hours of the second shift last from 2:45pm to 10:15pm. The starting and finishing times of the shifts are represented by the parameters $h_k^s$ and $h_k^f$, respectively, for $k = 1, 2$. The time unit is set to minutes. This implies that $h_1^s = 0$ and $h_1^f = 450$ for the first shift and $h_2^s = 390$ and $h_1^f = 840$ for the second shift. Each shift has a meal break of 30 min. The meal breaks must take place between 12:00pm and 12:30pm for the first shift and between 6:30pm and 7:00pm for the second shift. We use the parameters $b_k$ for the duration of the shift $k$. As a matter of fact, it is possible to use a single parameter for both shifts as they are equal. Parameters $b_1^s = 225$, $b_1^f = 255$ and $b_2^s = 615$, $b_2^f = 645$ denote the starting and finishing times of the first and second breaks, respectively.

Associated with all customer nodes, there exist earliest and latest service starting times which are denoted by parameters $e_i$ and $\ell_i$, respectively. It is worth noticing that these parameters designate the time interval during which service can begin at a customer node, but not the arrival time of the vehicle at the customer node. Whenever a vehicle arrives at the customer before $e_i$, then the vehicle has to wait for the delivery to take place until $e_i$. Actually, according to the company policy, a vehicle should not wait more than 5 min to serve a customer. The service time at a customer node is $s_i$ minutes for every $i \in N$. Among other parameters, we can mention parameter $t_{ij}$ that represents the traveling time from node $i \in N_0$ to node $j \in N_0$. Moreover, $c_{ij}$ denotes the cost of traveling from node $i \in N_0$ to node $j \in N_0$.

The decision variables are defined as follows. Binary variable $X_{ijk} = 1$ if a vehicle travels from node $i$ to node $j$ in shift $k$, it is zero otherwise. Note that we are not using an index for the vehicle since the vehicles are homogeneous. Another binary variable is used to determine the timing of meal breaks. Namely, $Y_{ik} = 1$ if a meal break is given by a vehicle in shift $k$ immediately after visiting node $i$, it is zero otherwise. There exist two continuous decision variables in the model. $T_{ik}$ denotes the starting time of the service at node $i$ in shift $k$ and $W_{ik}$ represents the waiting time of the vehicle before starting service at node $i$ in shift $k$. For the sake of clarity, we provide a table including all these elements with their explanations followed by model MBS1.

| Index sets and parameters | |
|---|---|
| $N$ | The set of customer (delivery) nodes |
| $N_0$ | The set of customer nodes and the depot |
| $K$ | The set of shifts |
| **Parameters** | |
| $c_{ij}$ | Cost of traveling from node $i \in N_0$ to node $j \in N_0$ |
| $t_{ij}$ | Traveling time from node $i \in N_0$ to node $j \in N_0$ |
| $e_i$ | Earliest time service can start at node $i$ |
| $\ell_i$ | Latest time service can start at node $i$ |
| $s_i$ | Service time for customer $i$ |
| $b_k$ | Duration of the meal break for shift $k$ |
| $b_k^s$ | Starting time of the meal break for shift $k$ |
| $b_k^f$ | Finishing time of the meal break for shift $k \in K$ |
| $h_k^s$ | Starting time of shift $k \in K$ |
| $h_k^f$ | Finishing time of shift $k \in K$ |
| $w_{\max}$ | Maximum time a vehicle is allowed to wait between arrival at a customer and starting service there |
| $M$ | A positive large number |
| **Decision variables** | |
| $X_{ijk}$ | 1 if a vehicle travels from node $i \in N_0$ to node $j \in N_0$ in shift $k \in K$; 0 otherwise |
| $Y_{ik}$ | 1 if a meal break for a vehicle in shift $k$ occurs after visiting node $i \in N$; 0 otherwise |
| $T_{ik}$ | Starting time of the service at node $i \in N$ in shift $k \in K$ |
| $W_{ik}$ | Waiting time of the vehicle before starting service at node $i$ in shift $k \in K$ |

Model MBS1:

$$\min V = \sum_{j \in N} \sum_{k \in \mathcal{K}} X_{0jk} \qquad (5.1)$$

subject to

$$\sum_{\substack{j \in N_0 \\ j \neq i}} \sum_{k \in \mathcal{K}} X_{ijk} = 1 \qquad\qquad i \in N \qquad (5.2)$$

$$\sum_{\substack{j \in N_0 \\ j \neq i}} \sum_{k \in \mathcal{K}} X_{jik} = 1 \qquad\qquad i \in N \qquad (5.3)$$

$$\sum_{\substack{j \in N_0 \\ j \neq i}} X_{ijk} = \sum_{\substack{j \in N_0 \\ j \neq i}} X_{jik} \qquad\qquad i \in N, k \in \mathcal{K} \qquad (5.4)$$

$$\sum_{k \in \mathcal{K}} X_{ijk} \leq 1 \qquad\qquad i \in N_0, j \in N_0 \qquad (5.5)$$

$$\sum_{i \in N} X_{0ik} = \sum_{i \in N} X_{i0k} \qquad\qquad k \in \mathcal{K} \qquad (5.6)$$

$$\sum_{k \in \mathcal{K}} Y_{ik} \leq 1 \qquad\qquad i \in N \qquad (5.7)$$

$$Y_{ik} \leq \sum_{j \in N_0} X_{ijk} \qquad\qquad i \in N, k \in \mathcal{K} \qquad (5.8)$$

$$\sum_{i \in N} Y_{ik} = \sum_{i \in N} X_{0ik} \qquad\qquad k \in \mathcal{K} \qquad (5.9)$$

$$e_i \leq T_{ik} \leq \ell_i \qquad\qquad i \in N, k \in \mathcal{K} \qquad (5.10)$$

$$T_{ik} + s_i + t_{ij} + b_k Y_{ik} \leq T_{jk} + M(1 - X_{ijk}) \qquad i \in N, j \in N, k \in \mathcal{K} \qquad (5.11)$$

$$T_{ik} + s_i \leq b_k^s + M(1 - Y_{ik}) \qquad\qquad i \in N, k \in \mathcal{K} \qquad (5.12)$$

$$T_{jk} \geq b_k^f Y_{ik} - M(1 - X_{ijk}) \qquad\qquad i \in N, j \in N, k \in \mathcal{K} \qquad (5.13)$$

$$T_{ik} + s_i + t_{i0} \leq h_k^f + M(1 - X_{i0k}) \qquad i \in N, k \in \mathcal{K} \qquad (5.14)$$

$$h_k^s + t_{0i} - M(1 - X_{0ik}) \leq T_{ik} \qquad\qquad i \in N, k \in \mathcal{K} \qquad (5.15)$$

$$W_{jk} \geq T_{jk} - (T_{ik} + s_i + t_{ij} + b_k Y_{ik}) - M(1 - X_{ijk}) \quad i \in N, j \in N, k \in \mathcal{K}$$
$$(5.16)$$

$$W_{ik} \leq w_{\max} \qquad\qquad\qquad\qquad\qquad\qquad i \in N, k \in \mathcal{K} \qquad (5.17)$$

$$X_{ijk} \in \{0, 1\} \qquad\qquad\qquad\qquad\qquad\qquad i \in N_0, j \in N_0, k \in \mathcal{K}$$
$$(5.18)$$

$$Y_{ik} \in \{0, 1\} \qquad\qquad\qquad\qquad\qquad\qquad i \in N, k \in \mathcal{K} \qquad (5.19)$$

$$T_{ik} \geq 0 \qquad\qquad\qquad\qquad\qquad\qquad\qquad i \in N, k \in \mathcal{K} \qquad (5.20)$$

$$W_{ik} \geq 0 \qquad\qquad\qquad\qquad\qquad\qquad\qquad i \in N, k \in \mathcal{K} \qquad (5.21)$$

In this formulation, (5.1) represents the objective function, which minimizes the number of vehicles used or equivalently the number of resulting routes. Constraints (5.2) imply that there is one departure from every customer node and it occurs in one of the shifts. By the same token, constraints (5.3) ensure that there is one arrival to every customer node and it occurs in one of the shifts. In other words, these two constraints guarantee that a customer is visited only once over two shifts. Constraints (5.4) are balance equations and ensure that if a customer is visited in a shift, then there must exist an incoming and outgoing arc to that customer. Constraints (5.5) make sure that an arc $(i, j)$ can either be traversed by a vehicle in one shift or not at all. Constraints (5.6) state that in a shift the number of vehicles leaving the depot must be equal to the number of vehicles entering the depot. Constraints (5.7) guarantee that a customer node can be selected for a meal break at most once. Constraints (5.8) imply that a meal break is possible at a customer node in a given shift if there is an outgoing arc from that node, i.e., that customer is visited. Constraints (5.9) state that the number of meal breaks in each shift must be as large as the number of vehicles in that shift. Constraints (5.10) ensure that the starting time of the service at a customer node is within the time windows determined by the earliest and latest times. Constraints (5.11) relate the service start times between two consecutive customer nodes. If customer $j$ is visited after customer $i$ by a vehicle, then the service start time at node $j$ cannot be earlier than the finish time at node $i$ given as $T_{ik} + s_i + t_{ij}$ delayed by the meal break, if any. If there is no travel from node $i$ to node $j$ in shift $k$, then $X_{ijk} = 0$ and constraint (5.11) becomes redundant. Constraints (5.12) imply that if a meal break in shift $k$ occurs after visiting node $i$ ($Y_{ik} = 1$), then the service at that node must be finished before the starting time of the break. To put it differently, customer node $i$ has to be the last node served before the meal break starts. Constraints (5.13) set a condition for the service start time at customer node $j$ that is visited right after the meal break. Namely, the starting time of the service at node $j$ cannot be earlier than the finishing time of the meal break if this node is visited immediately following node $i$ after which the break is given. Constraints (5.12) and (5.13) together do not allow more than one meal break on any route. Constraints (5.14) act as maximum tour duration constraints. If customer node $i$ is the last node visited of a tour in a shift, then $X_{i0k} = 1$ and the arrival time at the depot should not exceed the finishing time of the shift. In a similar vein, constraints (5.15) require that the starting time of the service at the customer immediately visited after the depot be later than the starting time of the shift. Constraints (5.16) are used to determine the waiting times at the customer nodes. If there is a travel from node $i$ to node $j$, then the waiting time at node $j$ is at least as large as the difference between the starting time of the service $T_{jk}$ at node $j$ and the arrival time at node $j$ given as $T_{ik} + s_i + t_{ij} + b_k Y_{ik}$. Note that, waiting time is not defined for the first customer of any route since the vehicle can adjust its departure time from the depot such that there will not be any waiting time at the first customer. Constraints (5.17) is an operational rule stated by the LSP dictating that the waiting time does not exceed $w_{max}$. Constraints (5.18) and (5.19) restrict $X_{ijk}$ and $Y_{ik}$ to be binary decision variables, respectively. Finally, constraints (5.20) and (5.21) define $T_{ik}$ and $W_{ik}$ as continuous variables.

The optimal solution of MBS1 provides the minimum number of routes (vehicles) $V^* = \sum_{j \in N} \sum_{k \in \mathcal{K}} X^*_{0jk}$ to have a feasible solution for a given problem instance without paying attention to the total distance traveled by the vehicles. Now, we make use of MBS2 which takes the number of routes as input and minimizes the total distance traveled, which we designate with $Z$. Hence, the mathematical programming formulation of MBS2 can be written as

Model MBS2:

$$\min Z = \sum_{i \in N_0} \sum_{j \in N_0} \sum_{k \in \mathcal{K}} c_{ij} X_{ijk} \tag{5.22}$$

subject to

$$\sum_{i \in N} \sum_{k \in \mathcal{K}} X_{0ik} \leq V^* \tag{5.23}$$

and constraints (5.2)–(5.21) $\tag{5.24}$

We remark that (5.23) is not redundant since the minimum total distance traveled may be obtained with a number of vehicles larger than $V^*$ in the absence of this constraints. Recall that the company policy dictates that the objective of minimizing the number of vehicles takes precedence over the objective of minimizing the distance.

## 5.4   The Proposed Heuristic TPH

This section explains the details of the heuristic proposed for solving VRPTW with shifts and meal breaks. The heuristic is composed of three main modules each of which has many sub-modules. In the first module, we generate an initial solution using a new route construction heuristic, and then improve it in terms of total distance traveled. The second module which is called Route Elimination is a search procedure that tries to reduce the number of routes. The solution given by this module is improved in the third module called Distance Minimization. The maximum waiting time between customers of a route constraint is maintained by forbidding any move that violates predefined maximum waiting time. In each module, procedure, or move, no modification is accepted that will lead to infeasible solution in terms of maximum waiting time. In the next subsections we will describe these modules first starting with the local search procedures employed within these modules.

### 5.4.1  Initial Solution Generation

Most of the route construction heuristics for VRPTW in the literature use an insertion criterion derived mainly as an extension of the insertion heuristic developed in [13] inspired by Clarke and Wright savings algorithm [1]. This insertion heuristic computes savings in total distance by regarding a limit for total waiting time of a route for feasibility while merging routes. According to the survey conducted by Bräysy and Gendreau [14], the most successful construction heuristic for VRPTW is I1 sequential heuristic developed by Ioannou et al. [15]. The insertion criterion is divided into three metrics to measure the impact of insertion of $i$. Those metrics are the impact of insertion of $i$ on itself (own impact), on the customers in the route that $i$ is inserted (internal impact), and on the non-routed customers (external impact). Total impact is calculated as a weighted combination of these impacts; and uses six different parameters. It requires parameter tuning and they could not offer a clear explanation of the relationships between the parameters and problem types. While I1 produces good solutions for randomly distributed customers having tight time windows, it causes excessive run times for the real case application in this paper where we have a larger number of randomly clustered customers, and wide time windows. Another well-known construction heuristic is the basic greedy heuristic which inserts the cheapest customer. However, it constructs too many routes in our case and thus turns out to be inefficient for the route elimination phase of our heuristic. Therefore, a new insertion heuristic referred to as Cost Ratio Based Insertion Heuristic (CRBI) is devised to construct an initial solution for VRPTW-BS.

The main decision to be made in VRPTW-BS besides the usual decisions for VRPs is assignments of customers to the shifts. Two shifts with overlapping times exist and there are some wide time windows such that most customers can be served in both shifts. To differentiate customers by their shifts and avoid a type incompatibility, each customer and each vehicle is assigned to a type which indicates their shifts and all moves are forbidden between incompatible types. Three types can occur in this problem:

- Type 0: The time window of a customer overlaps with both shifts or equivalently both shifts can serve it. This occurs if $e_i$ starts before the first shift ends ($e_i \leq h_1^f$) and $\ell_i$ is after the second shift starts ($\ell_i \geq h_2^s$) for customer $i$.
- Type 1: The time window of a customer overlaps with the first shift or equivalently only the first shift can serve it. This occurs if $\ell_i$ is before the second shift starts ($\ell_i \leq h_2^s$) for customer $i$.
- Type 2: The time window of a customer overlaps with the second shift or equivalently only the second shift can serve it. This occurs if $e_i$ is after the first shift ends ($e_i \geq h_1^f$) for customer $i$.

A vehicle can only serve in one shift. Its required departure time from the depot, arrival time at the depot, and meal break time are all specified by its shift. A vehicle is associated with a type determined by the dominating customer type among its customers. Type 1 and Type 2 dominate Type 0. If a vehicle has at least one Type 1 customer among its Type 0 customers, its type is designated as Type 1. The same is true for Type 2. After each modification in a vehicle's route, its type is updated. A vehicle is Type 0 only if all its customers are of Type 0. A Type 0 vehicle is feasible in terms of time windows, tour duration, capacity, and meal break if it can serve all the customers according to one of the shifts. It can be also said that the feasibility of a Type 0 vehicle is checked according to both shifts' specifications. If it is feasible in at least one of the shifts, it is accepted and its type is still 0. Its type is not updated as the defining shift type in which it is feasible just because all moves are performed according to type compatibility. A vehicle cannot have customers whose types are 1 and 2 simultaneously. Thus, Type 1 and Type 2 are incompatible in terms of shift. Any move between Type 1 customer and Type 2 customer, Type 1 vehicle and Type 2 vehicle, Type 1 vehicle and Type 2 customer, and Type 2 vehicle and Type 1 customer is forbidden due to type incompatibility.

While the insertion of a route is only possible at the end or beginning of another route in [1], our construction heuristic CRBI inserts a route into any position in another route when feasible and compatible. We remark that the sequence of nodes in route $u$ is preserved while inserting it into route $v$. To tackle the asymmetry in distances, reversals of routes are also considered. This means a route can be inserted into any position in another route after reversing its sequence.

For example when we want to insert *Route(u)*: 1-2 into *Route(v)*: 3-4, candidates of Route(u,v) after merging are obtained as follows: 1-2-3-4, 3-1-2-4, 3-4-1-2 - 2-1-3-4 - 3-2-1-4, 3-4-2-1, 1-2-4-3, 4-1-2-3, 4-3-1-2, 2-1-4-3, 4-2-1-3, and 4-3-2-1.

In addition to handling asymmetry, this insertion mechanism also helps to connect customers that are close spatially and temporally. For example, let us say that customers $i$ and $j$ are currently on different routes but spatially close to each other. Customer $i$ whose time window is wide is currently at the beginning of its route while earliest start time of $j$'s time window is late. In this case $j$ cannot be inserted next to $i$ because of time window infeasibility. However, reversal of the route that $i$ belongs to can be feasible allowing $i$ and $j$ to be connected successfully. At the beginning of CRBI, many insertions are possible while after merging routes one by one, possible insertions decrease dramatically since reversals and most of positions in other routes are not feasible due to time windows.

For each candidate $Route(u, v)$ generated as above, CRBI calculates a measure called $r(u, v)$, which represents the cost of the route per node visited. For each route pair $(u, v)$, the best insertion cost $r^*(u, v)$ is selected among all $r(u, v)$ values which represent the insertion costs. An $n \times n$ matrix called Ratio Matrix is created and maintained throughout the construction phase. The $u$th row and $v$th column of the Ratio Matrix contains $r^*(u, v)$. After all entries are calculated at the beginning of CRBI, some of them are updated as necessary during the subsequent iterations. Alternative cost definitions can be designed, which perform differently with respect to the problem characteristics such as the distribution of customers' locations, the length of scheduling period, tightness, and distribution of time windows. Solomon [13] created six different sets of benchmark instances regarding those factors. We defined the following three cost ratios and tested them on Solomon's benchmark instances as well as ours.

$$r_1(u, v) = \frac{Total\_distance}{Number\ of\ nodes\ in\ the\ candidate\ Route(u, v)} \tag{5.25}$$

$$r_2(u, v) = \frac{Total\_distance + Total\_waiting\_time}{Number\ of\ nodes\ in\ the\ candidate\ Route(u, v)} \tag{5.26}$$

$$r_3(u, v) = \frac{Tour\_duration - Service\_start\_time\_at\_first\_node}{Number\ of\ nodes\ in\ the\ candidate\ Route(u, v)} \tag{5.27}$$

$r_1(u, v)$ performs best when the instance contains clustered customers, while $r_2(u, v)$ is efficient when customer locations are uniformly distributed with some additional clusters. In our instances where there exist several distant customers besides a mix of uniformly distributed and clustered locations, $r_3(u, v)$ becomes the winner. Therefore, we use $r_3(u, v)$ in CRBI.

After calculating $r_3(u, v)$ for each candidate $Route(u, v)$, the best candidate $Route^*(u, v)$ with the minimum ratio $r_3^*(u, v)$ is selected and recorded as the entry in the $u$th row and $v$th column of a matrix. Some entries of this matrix may not exist due to the feasibility constraints in terms of time windows, route duration, and shift compatibility. The next step is to select the best $(u^*, v^*)$ pair with the smallest $r_3^*(u, v)$ to perform the insertion operation. CRBI heuristic is summarized in Algorithm 1.

| **Algorithm 1:** Cost ratio based insertion heuristic (CRBI) |
|---|

1: Input: Distance matrix, customer information including ID and Time Window, Shift information including shift starting and ending times, and meal break times

2: *Initialization:* Make $n$ routes: $0 \to i \to 0$ for each customer $i$. The type of each route is equivalent to the customer type it serves

3: *Ratio Matrix Construction:* Compute $r_3^*(u, v)$ for each $u$ and $v$ routes pair if feasible

4: **while** There exists at least one feasible entry in Ratio Matrix **do**

5:     *Best Insertion Decision:* Find the minimum $r_3^*(u, v)$ and perform the corresponding insertion operation. Assign vehicle type to newly created route with respect to dominant shift.

6:     *Update Ratio Matrix:* Delete all entries related to route $u^*$ and route $v^*$ from Ratio matrix, which are merged in step 5. Add a new row and column for the merged $Route(u^*, v^*)$ into Ratio matrix by computing $r(k, Route(u^*, v^*))$ and $r(Route(u^*, v^*), k)$ for each route $k$.

7: **end while**

8: **return** *Final routes*

As a final step, local search is applied to the initial solution generated by the CRBI heuristic. The moves implemented in local search are explained in the next subsection.

## 5.4.2  Local Search

Various local search (LS) moves are used within the proposed heuristic to improve the solution in terms of total distance traveled. Those procedures mainly take a solution and perform consecutive moves using first improvement strategy by obeying maximum waiting time constraint. They all use the same structure to select node(s) to search a cheaper solution and examine all nodes to modify if feasible. At the beginning, routes in the current solution are sorted with respect to nondecreasing distance per node, then the procedure continues by selecting a route starting from the route having the largest distance per node to the route having the smallest one. Following the route order, a predefined number of consecutive nodes of the selected route are examined to exchange with another predefined number of consecutive nodes in another route. For example, LS_SWAPXY tries to exchange $x$ consecutive nodes from the selected route with $y$ consecutive nodes from another route by examining all the routes until it finds a cheaper exchange. After searching the first improving swap for the $x$ many consecutive nodes and performing it if exists, the procedure continues with next consecutive $x$ many nodes from the selected route. If all consecutive $x$ many nodes from the route on hand are examined, next route having the largest distance per node is chosen to repeat the same search until all routes and all nodes are examined.

The move types used in the LS procedure are listed below:

- SWAPWITHINROUTE: Swaps two nodes' positions within the route.
- SWAP11: Swaps one node from a route with one node from another route.
- SWAP21: Swaps two nodes from a route with one node from another route.
- SWAP22: Swaps two nodes from a route with two nodes from another route.
- SWAP31: Swaps three nodes from a route with one node from another route.
- SWAP33: Swaps three node from a route with three nodes from another route.
- RELOCATION: Reinsert one node to another position in all routes including its original route.
- TAILEXCHANGE: Exchanges tails of two routes at any two positions.

The moves are applied to the initial solution generated by the CRBI heuristic in the following sequence: LS_SWAPWITHINROUTE, LS_SWAP11, LS_SWAP21, LS_SWAP22, LS_SWAP31, LS_SWAP33, and LS_TAILEXCHANGE. The reason for not using LS_RELOCATION is not to insert more nodes into smaller routes because it decreases the chance of eliminating them in the *Route Elimination* module.

## 5.4.3   Route Elimination

In this module, route elimination is achieved using two different procedures called *RelocationForRouteKilling* and *EjectionChain*. *RelocationForRouteKilling* works on a feasible solution where all nodes are served, while *EjectionChain* works on a partial solution where some nodes may not be served. These unserved nodes are kept in an ejection pool and they are tried to be inserted in a feasible way into existing routes with the aim of eliminating the route selected in the *RelocationForRouteKilling*. While reducing the number of routes, these procedures may deteriorate (increase) the total distance of the routes by forcing some nodes to be inserted into disadvantageous positions. Therefore, local search procedures are successively employed to arrange nodes in a way to reduce the total distance of the routes and create gaps between nodes to facilitate the moves in *RelocationFor-RouteKilling* or *EjectionChain* procedures that follow. Algorithm 2 summarizes the overall procedure.

### 5.4.3.1   RelocationForRouteKilling

*RelocationForRouteKilling* is a removal and insertion procedure that maintains feasibility. Selecting a route to eliminate and discovering the correct set of moves to eliminate it in a feasible fashion are difficult. This procedure considers several routes

---

**Algorithm 2:** Route elimination

---

1: Set $Iter = 0$
2: **while** $Iter < MaxIter1$ **and** (there is a change in the distance **or** the number of vehicles of the solution) **do**
3:     $Iter \leftarrow Iter + 1$
4:     Perform *RelocationForRouteKilling*
5:     Record the route $x$ with minimum number of nodes, remove it from the solution.
6:     Perform LS_SWAPWITHINROUTE, LS_SWAP11, LS_SWAP21, LS_SWAP22, LS_SWAP31, LS_SWAP33, LS_TAILEXCHANGE
7:     Perform *EjectionChain(route x)* to output *EjectionPool*
8:     **if** *EjectionPool* is not empty **then**
9:         Open new routes for the nodes in *EjectionPool* using CRBI
10:     **end if**
11:     Perform LS_SWAPWITHINROUTE, LS_SWAP11, LS_SWAP21, LS_SWAP22, LS_SWAP31, LS_SWAP33, LS_TAILEXCHANGE and LS_RELOCATION
12:     Update best solution
13: **end while**
14: **return** Best Solution

---

simultaneously for potential elimination. First, the routes in the current solution are sorted in the order of nondecreasing number of nodes. Starting from the smallest (first) route, each node in its sequence is tried to be inserted into other routes of the same or larger size if feasible regardless of the increase in the total distance. Whenever any node(s) in a route other than the first route is inserted somewhere else, the procedure starts over with a possibly new order of routes. The rationale behind this mechanism is to make use of the gap created in a relatively longer route in order to find more nodes to be relocated from shorter routes. Since it only searches over feasible insertion(s) limited to a given order of routes, the procedure gets stuck at a point. The procedure is presented in Algorithm 3.

The power of *RelocationForRouteKilling* is best understood when it is used with a naive construction heuristic such as the basic greedy construction heuristic. For example, basic greedy construction heuristic generates 68 routes for a large test instance with 1104 nodes. *RelocationForRouteKilling* takes this solution and reduces the number of routes from 68 to 41.

*RelocationForRouteKilling* might overlook possible route elimination opportunities since it allows insertions only into larger routes. This suggests a need for an extra insertion and removal procedure for route elimination. For this purpose, *EjectionChain* is implemented after *RelocationForRouteKilling*. But first, we remove the route having the smallest number of nodes and employ our local search procedures on the remaining routes to facilitate *EjectionChain* which aims to relocate the nodes of the removed route by a series of insertions and removals.

---

**Algorithm 3:** RelocationForRouteKilling

---

 1:  Set done = false
 2:  **while** not done **do**
 3:      Set done = true
 4:      Sort the routes in the order of nondecreasing number of nodes.
 5:      **for** Each Route in the sorted Route List  **do**
 6:          **for** Each Node in the selected route **do**
 7:              **for** Each Feasible insertion position in all routes of the same or larger size **do**
 8:                  Calculate the change in the total distance.
 9:              **end for**
10:          **end for**
11:      **end for**
12:      **if** There is feasible insertion(s) for the node on hand **then**
13:          Perform the best insertion.
14:          Set done = false
15:      **end if**
16:      **if** Insertion is performed for a node(s) from a route except the first route **then**
17:          exit for
18:      **end if**
19:      Perform LS_SWAPWITHINROUTE
20:  **end while**
21:  **return** Best Solution

---

### 5.4.3.2   EjectionChain

Ejection chains are proposed in [16] and have often been used in VRPTW problems. An ejection pool that contains unserved nodes is maintained throughout the procedure. By applying insertions and removals repeatedly, the nodes in the ejection pool are tried to be feasibly inserted into existing routes to eliminate the route given at the beginning of the procedure. *EjectionChain* works on partial solutions during or at the end of the procedure unless the ejection pool is empty. This application allows us to visit infeasible neighbors and increases diversification. Selection of nodes to eject in order to insert a node is another decision to make and many criteria are proposed to increase the efficiency of the procedure. These criteria mainly depend on relatedness of a node or a set of nodes to the node to be inserted. A relatedness criterion is measured with respect to spatial and temporal closeness. For detailed information, the reader may refer to the survey conducted in [17]. The *EjectionChain* procedure is summarized in Algorithm 4.

*EjectionChain* takes a route as input and initializes EjectionPool with the nodes in the given route. The nodes in the EjectionPool are tried to be inserted into remaining routes using *InsertFeasiblePositions* which uses the 2-regret insertion method. If all nodes are inserted, then the elimination of the given route is successful, and the procedure returns an empty EjectionPool with a feasible solution. If not, a node is selected in turn from the pool to be used in *InsertAndRemove* which is represented

---

**Algorithm 4:** EjectionChain (Route $r$)

---
1:   Initialize EjectionPool with given route $r$'s nodes.
2:   Perform *InsertFeasiblePositions* with EjectionPool
3:   **if** All unserved nodes in EjectionPool is inserted feasibly into other routes **then**
4:      **return** An empty EjectionPool
5:   **else**
6:      Iter=0
7:      **while** $Iter < MaxIter2$ and EjectionPool is not empty **do**
8:         Choose the first unserved node in EjectionPool and remove it from pool
9:         Perform *InsertAndRemove* for the selected node
10:        **if** the selected node is inserted feasibly by removing and inserting another node feasibly **then**
11:           Continue
12:        **else**
13:           Add removed node to EjectionPool
14:        **end if**
15:        $Iter \leftarrow Iter + 1$
16:      **end while**
17:   **end if**
18:   **return** EjectionPool

---

in Algorithm 5. This procedure simply searches a combination of feasible insertion of the given node by ejecting a node. It is successful if the ejected node is also inserted somewhere else. If it fails, the closest ejected node in terms of distance is selected and returned. The ejected node is added to the EjectionPool. Relatedness is defined in terms of distance for two reasons. First, it inserts a closer node to the modified route and the increase in its distance is smaller, since it serves originally the ejected node which does not differ much in terms of location. Secondly, ejecting the closest node preserves the closeness of the unserved nodes in the EjectionPool, which will be easier to construct only a route for nodes in EjectionPool when the procedure ends with a non-empty EjectionPool. By doing so, the route elimination iteration is less likely to destroy a route and yield more than a route. Inserting and removing procedure is repeated until a predefined number of iterations are reached or EjectionPool becomes empty. It returns EjectionPool which might be empty or contain unserved nodes.

*EjectionChain* procedure returns a partial solution if the EjectionPool is not empty. Route construction heuristic CRBI is used to generate route(s) for the nodes in the EjectionPool to have a feasible solution where all nodes are served. After restoring the feasibility, local search procedures are applied to the solution to improve the routes in terms of distance. Successive applications of *RelocationFor-RouteKilling* and *EjectionChain* reduce the total number of routes if possible for a predefined number of iterations or time limit.

---

**Algorithm 5:** InsertAndRemove (Node *n*)

---

 1: **for** Each Route in the current solution **do**
 2:   **for** Each position in the route on hand **do**
 3:     Insert node *n* into the selected position
 4:     **if** The insertion is feasible **then**
 5:       **return** no ejected node
 6:     **else**
 7:       **for** Each node up to the first infeasible node **do**
 8:         Remove the selected node
 9:         **if** The route is feasible after the removal of the node **then**
10:           Perform *InsertFeasiblePositions* with removed node
11:           **if** The removed node is inserted in a feasible way **then**
12:             **return** no ejected node
13:           **else**
14:             Calculate the closeness between the removed node and node *n*
15:             **if** It is the closest one found so far **then**
16:               Save the route with corresponding insertion and removal details
17:             **end if**
18:             Insert the removed node back to its position
19:           **end if**
20:         **end if**
21:       **end for**
22:       Restore the original version of the route on hand
23:     **end if**
24:   **end for**
25: **end for**
26: **if** A close ejected node is found in Line 15 **then**
27:   Perform corresponding insert and remove operation for node *n* and the closest ejected node
28:   **return** the closest ejected node
29: **else**
30:   **return** node *n*
31: **end if**

---

## 5.4.4  Distance Minimization

Up to this point, the main concern was to reduce the number of routes, whereas improving the total distance was the second concern. In this section given the best solution obtained by *RouteElimination* module, local search procedures are performed to improve the routes in terms of total distance. LS_SWAPWITHINROUTE, LS_SWAP11, LS_SWAP21, LS_SWAP22, LS_SWAP31, LS_SWAP33, LS_RELOCATION, and LS_TAILEXCHANGE are iteratively applied in the given order for a predefined number of iterations or until no change occurs in the current solution. Figure 5.2 depicts the overall heuristic implemented in this paper.

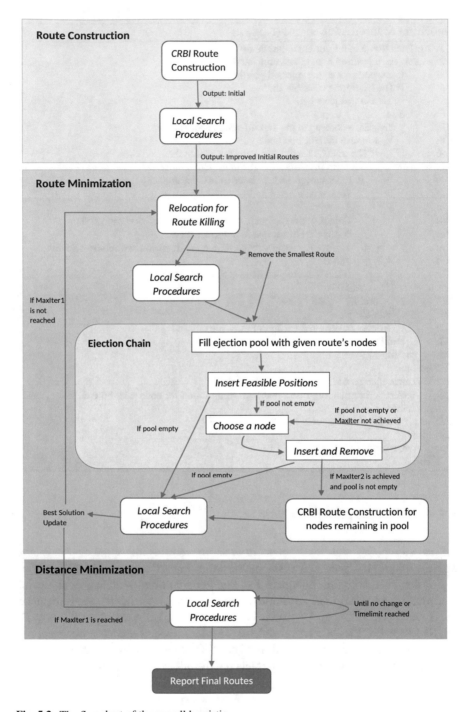

**Fig. 5.2** The flowchart of the overall heuristic

## 5.5  Computational Results

Since the VRPTW variant explored in this paper, VRPTW-BS, has not been studied
before in the literature, there is no benchmark instance to assess the quality of the
solutions produced by our heuristic. We obtain several real-life instances from the
logistics providing company. However, to disclose the data of the company such as
the number of customers served in a day, service times, time windows of customers,
neither the values of these parameters nor the results of the real instances are given in
this section. Instead, two different sets of test instances are created based on the real
data corresponding to different days of operation. Those test instances are generated
by selecting random subsets of customers while maintaining problem structure in
terms of distribution of locations, time windows' lengths, the distribution of time
windows among customers, shift specifications, and service times. The first set is
composed of smaller instances which are solved using the mathematical models
formulated in Sect. 5.3 for assessing the quality of the solutions obtained by the
TPH. The second set of larger test instances is created to compare the running times
and solution quality with respect to another method based on cluster-first, route-
second approach. The values of the parameters $MaxIter1$ and $MaxIter2$ are set to
10 and 100, respectively. The proposed heuristic TPH is coded in C# and executed
in a computer equipped with Intel® CoreTM i7 CPU @ 3.07 GHz processor and 8
GB RAM and Windows 10 64-bit operating system. The MILP models are solved
using the callable library of IBM ILOG CPLEX Optimization Studio V12.8.0.

### 5.5.1  Computational Results on Small Test Instances

The sizes of the small test instances vary from 10 to 50 nodes in the increment of
five nodes, which gives rise to 18 instances with two instances for each size. All
small test instances are firstly solved for minimizing the number of vehicles as the
objective under a given time limit of 6 h. Then, they are solved for minimizing the
total distance traveled for the fixed number of vehicles found as the best feasible
solution in the first phase. The instances having two vehicles in their best solutions
are optimal in terms of the number of vehicles since they all have Type 1 and Type
2 customers and the lower bound on the number of vehicles is two, i.e. a vehicle for
each shift.

    Table 5.1 includes the results obtained by the mathematical models as well as
TPH. Namely, the number of routes (or equivalently the number of vehicles) given
by MBS1 ($V_{MBS1}$) and the total distance provided by MBS2 ($Z_{MBS2}$) corresponding
to $V_{MBS1}$. The second column of the table indicates the number of customer nodes
in the instance, and an asterisk next to the number of routes means that MBS1 could
not produce the optimal solution. Moreover, "Opt. Gap" represents the percent gap
between the best feasible solution and the best possible solution obtained by the
mathematical model MBS2 for the total distance, and lastly $PD$ denotes the percent

**Table 5.1** Results on small test instances

| Instance | $|N|$ | Best solution given by MILPs | | | | Best solution given by TPH | | | |
|---|---|---|---|---|---|---|---|---|---|
| | | $V_{MBS1}$ | $Z_{MBS2}$ | Opt. Gap (%) | Run time (s) | $V_{TPH}$ | $Z_{TPH}$ | Run time (s) | PD (%) |
| S1 | 10 | 2 | 104.29 | 0.00 | 0.33 | 2 | 104.60 | 0.15 | 0.30 |
| S2 | 10 | 2 | 121.94 | 0.00 | 0.24 | 2 | 121.94 | 0.02 | 0.00 |
| S3 | 15 | 2 | 81.94 | 0.00 | 10.33 | 2 | 83.38 | 0.46 | 1.73 |
| S4 | 15 | 1 | 137.54 | 0.00 | 0.20 | 1 | 142.97 | 0.17 | 3.80 |
| S5 | 20 | 2 | 140.17 | 0.00 | 1.64 | 2 | 143.86 | 0.33 | 2.57 |
| S6 | 20 | 2 | 133.02 | 0.00 | 1570.92 | 2 | 134.91 | 1.12 | 1.40 |
| S7 | 25 | 3 | 283.12 | 0.00 | 94.67 | 3 | 285.56 | 0.76 | 0.85 |
| S8 | 25 | 2 | 188.64 | 0.00 | 33.33 | 2 | 188.76 | 0.84 | 0.07 |
| S9 | 30 | 2 | 101.63 | 6.00 | 21600 | 2 | 102.06 | 0.91 | 0.41 |
| S10 | 30 | 3* | 229.50 | 3.42 | 21600 | 3 | 233.39 | 0.84 | 1.67 |
| S11 | 35 | 3* | 246.44 | 6.55 | 21600 | 3 | 268.01 | 2.18 | 8.05 |
| S12 | 35 | 2 | 177.67 | 4.83 | 21600 | 2 | 177.36 | 9.10 | −0.17 |
| S13 | 40 | 2 | 226.36 | 25.58 | 21600 | 2 | 244.91 | 8.81 | 7.58 |
| S14 | 40 | 2 | 278.37 | 18.32 | 21600 | 2 | 294.53 | 1.45 | 5.49 |
| S15 | 45 | 3* | 259.21 | 6.20 | 21600 | 3 | 270.82 | 2.46 | 4.29 |
| S16 | 45 | 2 | 144.53 | 10.93 | 21600 | 2 | 151.29 | 14.27 | 4.47 |
| S17 | 50 | 3* | 319.01 | 31.30 | 21600 | 3 | 326.56 | 3.17 | 2.31 |
| S18 | 50 | 3* | 337.23 | 10.83 | 21600 | 2 | 338.55 | 8.33 | 0.39 |

deviation of the total distance produced by the TPH from that given by MBS2 computed as $100 \times (Z_{TPH} - Z_{MBS2})/Z_{MBS2}$.

The proposed heuristic is successful at minimizing the number of routes for 18 small test instances while maintaining a percent deviation of 2.51% on average in terms of the total distance. We also note that on eight small instances that can be solved to optimality by the mathematical model, i.e., instances with 25 customers and less, the average PD turns out to be 1.34%. Also, MBS1 could not attain a feasible solution with less than three vehicles in 21600 s for S18, while the TPH finds a better solution having two vehicles in 8.33 s. Recall that the number of routes given by the mathematical model MBS1 is not optimal for instances S10, S11, S15, S17, and S18.

## 5.5.2 Computational Results on Large Test Instances

In order to evaluate the quality of the solutions generated by our heuristic, we implement a method that is inspired by the traditional method employed by the company. The company used to divide its customers into clusters with respect to predefined geographic regions and determine a single route for each cluster using a variant of the nearest insertion heuristic. This was possible as the number of customers in each cluster was small enough to be served by a vehicle. However, this method cannot be used as a benchmark since it does not consider meal breaks, shifts, and time windows constraints. Therefore, we extend it by incorporating dynamic clustering into our heuristic. This means that the customers in the original problem instance are partitioned into a number of clusters by using $k$-means clustering method first proposed by MacQueen [18] based on geographic closeness and our heuristic is applied to each of those clusters, separately.

In Table 5.2 we present the results of different number of clusters used in $k$-means clustering in order to see its effect on the performance. The second column of the table indicates the number of customer nodes in the instance, and each of the following three blocks corresponds to a different number of clusters, i.e. four, six, and 10 clusters. In each block, the first column represents the number of routes (vehicles), the second column gives the total distance traveled, and the third column provides the computation time. The first observation we can make based on the results in Table 5.2 is that more clusters result in more routes in most of the test instances. Only three out of 13 instances have better solutions in case of using six clusters rather than four clusters. Also, dividing the customer nodes into ten clusters yields worse solutions in terms of the total number of routes for all instances.

In Table 5.3, we present the results obtained for larger instances by the TPH without clustering the customers and compare them to the best results given by the cluster-first, route-second approach using $k$-means clustering. The first group of columns contains the results given by TPH without clustering. Within this group, the number of routes ($V$), the total distances traveled ($Z$), and computation time can be seen for both the initial solution obtained using CRBI and the final solution given by TPH. The second group of columns displays the best solutions attained by TPH with clustering for each instance. The last groups of columns show the relative quality of

Table 5.2 $k$-means results on large test instances

| Instance | $|N|$ | $k=4$ | | | $k=6$ | | | $k=10$ | | |
|---|---|---|---|---|---|---|---|---|---|---|
| | | V | Z | Time (min) | V | Z | Time (min) | V | Z | Time (min) |
| L1 | 587 | 20 | 2241.36 | 1.63 | **20** | **2207.15** | **2.27** | 29 | 2458.92 | 0.51 |
| L2 | 735 | 26 | 2650.83 | 2.74 | **23** | **2560.92** | **3.30** | 29 | 2630.34 | 1.16 |
| L3 | 806 | 24 | **2717.46** | 4.70 | 26 | 2716.33 | 2.36 | 29 | 2694.10 | 2.52 |
| L4 | 834 | **26** | **2892.23** | **4.36** | 27 | 2882.47 | 5.42 | 29 | 2972.42 | 2.75 |
| L5 | 855 | **26** | **2929.89** | **4.72** | 28 | 2905.56 | 3.91 | 31 | 3034.79 | 3.44 |
| L6 | 856 | **26** | **2803.24** | **3.53** | 29 | 2838.99 | 2.26 | 30 | 2865.24 | 2.26 |
| L7 | 1064 | 38 | 4161.77 | 6.25 | **37** | **4257.25** | **4.27** | 41 | 4233.89 | 3.73 |
| L8 | 1104 | 36 | **4088.66** | **7.13** | 37 | 4206.64 | 5.08 | 42 | 4301.38 | 3.96 |
| L9 | 1232 | 39 | **4483.63** | **8.58** | 39 | 4500.11 | 6.65 | 44 | 4627.63 | 3.68 |
| L10 | 1254 | 38 | **4536.33** | **15.04** | 40 | 4601.52 | 9.42 | 42 | 4591.47 | 6.06 |
| L11 | 1258 | 38 | **4286.33** | **9.55** | 42 | 4346.89 | 5.88 | 44 | 4430.56 | 6.26 |
| L12 | 1259 | 36 | **3905.71** | **8.25** | 37 | 3839.87 | 10.16 | 41 | 3874.27 | 10.74 |
| L13 | 1307 | **40** | **4544.30** | **10.63** | 41 | 4512.45 | 8.37 | 44 | 4645.77 | 6.77 |

**Table 5.3** Comparison of methods

| | TPH without Clustering | | | | | | TPH with Clustering | | | | Effect of Clustering | | | |
| | Initial Solution | | | Final Solution | | | Best Solution | | | | Initial Solution | | Final Solution | |
| Instance | V | Z | Time (min) | V | Z | Time (min) | # Clus. | V | Z | Time (min) | Incr. in V | % Incr. in Z | Incr. in V | % Incr. in Z |
|---|---|---|---|---|---|---|---|---|---|---|---|---|---|---|
| L1 | 19 | 2546.25 | 0.51 | 18 | 2158.33 | 5.46 | 6 | 20 | 2207.15 | 2.27 | 1 | −15.36 | 2 | 2.21 |
| L2 | 24 | 2926.37 | 0.89 | 22 | 2497.13 | 8.2 | 6 | 23 | 2560.92 | 3.3 | −1 | −14.27 | 1 | 2.49 |
| L3 | 25 | 3148.49 | 1.14 | 23 | 2698.99 | 10.31 | 4 | 24 | 2717.46 | 4.7 | −1 | −15.86 | 1 | 0.68 |
| L4 | 25 | 3245.02 | 1.12 | 24 | 2895.28 | 10.74 | 4 | 26 | 2892.23 | 4.36 | 1 | −12.2 | 2 | −0.11 |
| L5 | 27 | 3393.46 | 1.26 | 25 | 2909.7 | 12.7 | 4 | 26 | 2929.89 | 4.72 | −1 | −15.82 | 1 | 0.69 |
| L6 | 28 | 3231.46 | 1.16 | 25 | 2759.05 | 11.27 | 4 | 26 | 2803.24 | 3.53 | −2 | −15.28 | 1 | 1.58 |
| L7 | 35 | 4759.84 | 1.81 | 32 | 4101.39 | 17.54 | 6 | 37 | 4257.25 | 4.27 | 2 | −11.81 | 5 | 3.66 |
| L8 | 37 | 4891.18 | 2.02 | 33 | 4119.44 | 17.41 | 4 | 36 | 4088.66 | 7.13 | −1 | −19.63 | 3 | −0.75 |
| L9 | 40 | 5384.49 | 2.56 | 37 | 4375.31 | 24.83 | 4 | 39 | 4483.63 | 8.58 | −1 | −20.09 | 2 | 2.42 |
| L10 | 41 | 5551.42 | 2.54 | 37 | 4671.81 | 21.66 | 4 | 38 | 4536.33 | 15.04 | −3 | −22.38 | 1 | −2.99 |
| L11 | 40 | 4960.89 | 2.84 | 37 | 4294.02 | 24.24 | 4 | 38 | 4286.33 | 9.55 | −2 | −15.74 | 1 | −0.18 |
| L12 | 38 | 4400.32 | 2.9 | 35 | 3819.6 | 26.6 | 4 | 36 | 3905.71 | 8.25 | −2 | −12.66 | 1 | 2.2 |
| L13 | 40 | 5276.61 | 2.89 | 39 | 4656.17 | 27.86 | 4 | 40 | 4544.30 | 10.63 | 0 | −16.11 | 1 | −2.46 |

the solutions given by the TPH with clustering with respect to the solutions of the TPH without clustering in terms of two objective values. The columns titled "Incr. in $V$" and "% Incr. in $Z$" represent, respectively, the increase in the number of routes and the percent increase in the number for total distance traveled when TPH with clustering is employed.

In four out of 13 instances, i.e., instances L1, L4, L7, and L13, initial route construction without clustering achieves less or equal number of routes than the best $k$-means results while the total distances are 15.94% worse on average than the best solutions of $k$-means. The final solutions of the TPH without clustering are always superior to the best solutions of the TPH with clustering in terms of the number of routes which is the primary objective of the problem. In terms of total traveling distance, TPH without clustering is again better than its counterpart since it yields a total traveling distance that is 0.73% shorter on the average. Having a smaller number of routes does not necessarily yield shorter total traveling distances as we can see in the case of problem instance L13. Observe that the number of routes is 40 when clustering is performed with a total distance of 4544.30, while TPH without clustering provides 39 routes with a total distance of 4656.17. Since the primary performance measure (objective function) of the LSP is route minimization, the latter solution is better.

## 5.6 Conclusion and Future Remarks

In this study we address an extension of the classical Vehicle Routing Problem with Time Windows, which incorporates shifts as well as meal breaks of the delivery personnel. There is also an additional condition that requires the waiting time of a vehicle occurring between the customer arrival time and service start time to be less than a certain threshold value. For the solution of this problem called the Vehicle Routing Problem with Time Windows including Meal Breaks and Shifts (VRPTW-BS) we develop three methods. The first method that can be implemented for small-sized instances with up to 35 customers is a mixed-integer linear programming model that can be solved by commercial solvers such as CPLEX and Gurobi with optimality gaps less than 7%. Notice that there do not exist published works in the literature which report feasible solutions for Vehicle Routing Problem with Time Windows including lunch breaks for 35 nodes. The second method is based on a two-phase heuristic solution procedure called TPH that can be utilized for larger instances with more than 1000 customers. The first phase of TPH consists of a construction heuristic that generates a feasible solution with respect to all the restrictions imposed. This solution is further improved in a second post-optimization phase by two different local search procedures implemented with several move operators separately for route elimination and distance minimization. The third method is to partition the customers into non-overlapping clusters in terms of the geographical locations by means of the well-known $k$-means clustering algorithm, and then applying our two-phase heuristic to the customer set of each cluster. Obviously, the third approach is the fastest method, as the number of customers becomes smaller in each cluster. But this advantage comes at the expense of reduced solution quality. Consequently, for large instances the best performing approach turns out to be the two-phase heuristic without clustering. Based on the assessment of this heuristic on eight small instances that can be solved to optimality by the mathematical model, we conclude that it performs quite well with an average percent deviation of 1.34%. On all 18 small instances, the percent deviation turns out to be 2.51%.

There exist several directions in which this study can be further extended. One of them is the incorporation of pick-up customers into the set of delivery customers. This means that vehicles need to visit in their routes both pick-up customers with time windows as well as delivery customers with time windows. This extension, however, dictates that the assumption of uncapacitated vehicles cannot be made anymore since the available space in the vehicles has to be considered before visiting pick-up customers. Another extension is more challenging and involves the dynamic construction of the routes and making price discrimination for customers. In other words, depending on the availability within the routes, some customers may be more easily accommodated into the routes than others, which may help in offering lower prices to some customers.

It could also be interesting to devote some research effort for the investigation of partitioning customers into clusters taking into account not only spatial characteristics but also temporal ones of the customers. Clustering becomes a nontrivial problem in the presence of time windows and especially when there exist shifts because the distance measure between the customer nodes should consider their spatial and temporal attributes. In other words, closeness of the customers must be defined with respect to both geographic position as well as time windows even though most distance measures are based on geographic information. The problem specified in this paper adds another dimension, which is the shift information, making the clustering even more difficult.

# References

1. Clarke, G., Wright, J.W.: Scheduling of vehicles from a central depot to a number of delivery points. Oper. Res. **12**(4), 568–581 (1964)
2. Coelho, L.C., Gagliardi, J.P., Renaud, J., Ruiz, A.: Solving the vehicle routing problem with lunch break arising in the furniture delivery industry. J. Oper. Res. Soc. **67**(5), 743–751 (2016)
3. Toth, P., Vigo, D.: Vehicle Routing, Problems, Methods and Applications, 2nd edn. Society for Industrial and Applied Mathematics, Philadelphia (2014)
4. Buhrkal, K., Larsen, A., Ropke, S.: The waste collection vehicle routing problem with time windows in a city logistics context. Procedia Soc. Behav. Sci. **39**, 241–254 (2012)
5. Sahoo, S., Kim, S., Kim, B.-I., Kraas, B., Popov, A.: Routing optimization for waste management. Interfaces **35**, 24–36 (2005)
6. Taillard, E., Badeau, P., Gendreau, M., Guertin, F., Potvin, J.Y.: A tabu search heuristic for the vehicle routing problem with soft time windows. Transp. Sci. **31**(1), 170–186 (1997)

7. Sze, S.-N.: A study on the multi-trip vehicle routing problem with time windows and meal break considerations. A thesis submitted in partial fulfillment of the requirements for the degree of Doctor of Philosophy, University of Sydney (2011)
8. Kim, B.-I., Kim, S., Sahoo, S.: Waste collection vehicle routing problem with time windows. Comput. Oper. Res. **33**(12), 3624–3642 (2006)
9. Benjamin, A.M., Beasley, J.E.: Metaheuristics for the waste collection vehicle routing problem with time windows, driver rest period and multiple disposal facilities. Comput. Oper. Res. **37**(12), 2270–2280 (2010)
10. Vidal, T., Crainic, T.G., Gesndreau, M., Prins, C.: A unified solution framework for multi-attribute vehicle routing problems. Eur. J. Oper. Res. **234**(3), 658–673 (2014)
11. Markov, I., Varone, S., Bierlaire, M.:x Integrating a heterogeneous fixed fleet and flexible assignment of destination depots in the waste collection VRP with intermediate facilities. Transp. Res. B Methodol. **84**, 256–274 (2019)
12. Keskin, M., Laporte, G., Çatay, B.: Electric vehicle routing problem with time-dependent waiting times at recharging stations. Comput. Oper. Res. **107**, 77–94 (2019)
13. Solomon, M.M.: Algorithms for the vehicle routing and scheduling problems with time window constraints. Oper. Res. **35**(2), 254–265 (1987)
14. Bräysy, O., Gendreau, M.: Vehicle routing problem with time windows, Part I: Route construction and local search algorithms. Transp. Sci. **39**(1), 104–118 (2005)
15. Ioannou, G., Kritikos, M., Prastacos, G.: A greedy look-ahead heuristic for the vehicle routing problem with time windows. J. Oper. Res. Soc. **52**(5), 523–537 (2001)
16. Glover, F.: New ejection chain and alternating path methods for traveling salesman problems. In: Computer Science and Operations Research (pp. 491–509). Pergamon (1992)
17. Solak, S.: Advanced Multi-Stage Local Search Applications to Vehicle Routing Problem with Time Windows: A Review. Technical Report (2005)
18. MacQueen, J.: Some methods for classification and analysis of multivariate observations. In: Proceedings of the Fifth Berkeley Symposium on Mathematical Statistics and Probability, Vol. 1(14), pp. 281–297 (1967)

# Chapter 6
# A Decomposition-Based Heuristic for a Waste Cooking Oil Collection Problem

Ceren Gultekin, Omer Berk Olmez, Burcu Balcik, Ali Ekici, and Okan Orsan Ozener

**Abstract** Every year, a tremendous amount of waste cooking oil (WCO) is produced by households and commercial organizations, which poses a serious threat to the environment if disposed improperly. While businesses such as hotels and restaurants usually need to have a contract for their WCO being collected and used as a raw material for biodiesel production, such an obligation may not exist for households. In this study, we focus on designing a WCO collection network, which involves a biodiesel facility, a set of collection centers (CCs), and source points (SPs) each of whom represents a group of households. The proposed location-routing problem (LRP) determines: (i) the CCs to be opened, (ii) the number of bins to place at each CC, (iii) the assignment of each SP to one of the accessible CCs, and (iv) the vehicle routes to collect the accumulated oil from the CCs. We formulate the problem as a mixed-integer mathematical model and solve it by using commercial solvers by setting a 1-h time limit. We also propose a decomposition-based heuristic and conduct a computational study. Our decomposition algorithm obtains the same or better solutions in 95% of all the test instances compared to the proposed mathematical model.

## 6.1 Introduction

Huge amounts of waste cooking oil (WCO) are generated by the food service industry (such as restaurants, hotels, canteens, fast food chains, etc.) and also by households during food processing, production, and cooking. Inappropriate disposal of WCO may cause significant harm to the environment and also the pipeline infrastructure. Specifically, at low temperatures the WCO that is poured down the sink or toilet may congeal in the drain pipes, clog the sewage system, and restrict

C. Gultekin (✉) · O. B. Olmez · B. Balcik · A. Ekici · O. O. Ozener
Department of Industrial Engineering, Ozyegin University, Istanbul, Turkey
e-mail: ceren.gultekin.4430@ozu.edu.tr; berk.olmez@ozu.edu.tr; burcu.balcik@ozyegin.edu.tr; ali.ekici@ozyegin.edu.tr; orsan.ozener@ozyegin.edu.tr

© Springer Nature Switzerland AG 2020
H. Derbel et al. (eds.), *Green Transportation and New Advances in Vehicle Routing Problems*, https://doi.org/10.1007/978-3-030-45312-1_6

the wastewater flow. If the WCO is directly thrown into garbage bins, it may cause fires at landfill sites or rubbish dumps. It may also leak into soil and mix into groundwater, which is one of the main sources for drinking water. In addition to physical damages due to inappropriate disposal of WCO such as contamination in water reservoirs and soil, the cost incurred for wastewater treatment and infrastructure repairment is substantial as well. Due to the considerable increase in food consumption and subsequent WCO production, an annual amount of 4.5–11.3 million liters of WCO occurs in the USA together with an amount of $4 \times 10^5$–$6 \times 10^5$ tons of WCO in Japan [1]. The approaches followed to manage this high amount of WCO carry significant importance to avoid the already mentioned environmental and infrastructural damages.

A sustainable alternative for appropriate disposal is collecting and using the WCO as a raw material in biodiesel production plants. Recently, biodiesel production and usage are gaining more popularity since the environmental concerns about the non-renewable energy sources lead authorities to focus more on substitutes for petroleum products. It is claimed that the global supply of petroleum and natural gas from conventional sources will fail to satisfy the growing demand for energy over the next 25 years [2]. The environmental benefits that bring out biodiesel as an attractive substitute for petrodiesel are basically due to its high biodegradability and its natural origin derived from renewable biomass sources. Biodiesel degrades about four times faster than petroleum diesel [3], and hence it is a more eco-friendly energy source. Therefore, promoting the use of WCO as a biodiesel source has the potential to reduce $CO_2$, particle pollution, and other greenhouse gases as the carbon contained in biomass-derived fuel is largely biogenic and renewable [4]. Unlike, diesel is one of the major vehicle fuels that produce black smoke particles and $SO_2$ emissions; it is indeed responsible for one-third of the total transport generated greenhouse gas emissions [5]. Gonzalez-Gomez et al. [6] conduct a series of tests on a diesel engine to compare the exhaust emissions and performance characteristics of vegetable-based WCO methyl ester and mineral diesel fuel. The results show that using WCO methyl ester produces considerably lower levels of CO, $CO_2$, smoke (approximately 64%, 7.5%, and 48%, respectively), and $SO_2$.

The general process of biodiesel production by using WCO occurs in four stages: (1) collection, (2) pre-treatment, (3) delivery, (4) transesterification. The WCO is first collected from food service sector locations and households by authorized biodiesel companies. After collection, it is pre-treated to remove the solid particles and water to improve its quality. The cleansed oil is loaded in cistern trucks and delivered to the biodiesel facility to be then transesterified with methanol to biodiesel [7]. Transesterification of vegetable oils and animal fats constitutes the most common method for biodiesel production among the other three primary processes; direct use and blending, microemulsions, and thermal cracking (pyrolysis) [8]. However, the use of virgin oils for this process imposes a significant manufacturing cost such that biodiesel production becomes 1.5 times more expensive than petroleum-based diesel production. Nevertheless, 70–95% of the overall biodiesel production cost is issued due to the cost of raw material, which is vegetable oil or animal fats [9]. WCO represents a much less expensive substitute for pure vegetable oil; therefore, it is crucial to promote effective collection of WCO.

In the European food service sector, it is estimated that 700,000 tons of WCO are being collected per year while the potential lies around 830,000 tons. At household level, it is estimated that less than 50,000 tons of WCO are collected yearly while there is a potential to increase this amount to about 800,000–900,000 tons (https:// www.theicct.org/sites/default/files/publications/Greenea%20Report%20Household %20UCO%20Collection%20in%20the%20EU_ICCT_20160629.pdf). The reason that there is a greater WCO collection effort in the commercial sector is primarily due to laws and regulations, whereas household collection may be affected by various factors such as recycling habits, urbanization, incentives, and policies, etc. WCO collection through households has not yet matured in Europe, which constitutes an opportunity area for reaching the 2030 targets proposed by the European Commission in Renewable Energy Directive (RED II) (https://eur-lex. europa.eu/legal-content/EN/TXT/?uri=CELEX%3A52016PC0767R%2801%29). Specifically, in this proposal, a target has been set to 27% for the share of renewable energy consumed in Europe in 2030, while the actual share was measured as 17.5% in 2017 (https://ec.europa.eu/eurostat/statistics-explained/index.php/Renewable_ energy_statistics#Renewable_energy_produced_in_the_EU_increased_by_two_ thirds_in_2007-2017).

In this study, we focus on designing a WCO collection system, which will support effective and efficient collection of WCO from the households. In particular, we are motivated by the oil collection operations of a biodiesel company based in Istanbul/Turkey, which is described by Karabak [10]. In this oil collection network, the biodiesel company collects the WCO accumulated at businesses operating in the food service sector and already has an established logistical system for visiting the restaurants, hotels, etc. However, it is more challenging to collect oil from households and individuals due to various factors. Primarily, businesses such as hotels and restaurants have contracts, obliged by laws and regulations, for disposing their WCO, whereas such an obligation does not exist for households. Therefore, environmental awareness of the individuals and society, in general, plays a large role in disposing WCO properly. Moreover, due to the large quantity of households, on-demand collection from them becomes an inefficient approach. Hence, households need to access to the selected collection centers by their own means, while businesses such as restaurants are visited by the vehicles of the authorized biodiesel companies. This situation necessitates higher number of collection centers (CCs) to be placed for people to easily reach and bring their WCO. In this study, we propose a mathematical model and a solution method to design a WCO collection network. Specifically, we consider a network that involves a set of households which are referred as source points (SPs), a set of CCs, and a depot, which represents the biodiesel production facility where we assume the pre-treatment phase is handled as well. An example network representing our problem setting is illustrated in Fig. 6.1. There are four decisions to consider in this problem: (1) Location decisions; which CCs should be opened, (2) Allocation decisions; how many bins should be placed at each CC, (3) Assignment decisions; which SP should be assigned to which CC,

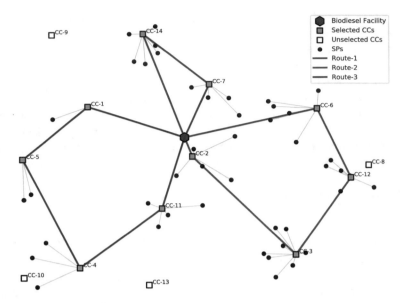

**Fig. 6.1** An example WCO collection network

(4) Routing decisions; the order of visits to the selected CCs during the collection days. We assume that the collections are made periodically (i.e., once per week). The objective of this problem is to minimize the total cost, which includes fixed costs for locating CCs, costs of bins to be placed at the CCs, and transportation costs.

The proposed problem is a variant of the secondary-facility location-routing problem (LRP), which assumes that the primary facilities (plants) are located at known and fixed sites and the facility location decisions are made upon the secondary facilities (intermediate facilities) [11]. In our problem, the location of the biodiesel facility (plant) is also fixed and known, whereas the location decisions are made for the CCs (intermediate facilities). However, our problem differs from secondary-facility LRP in some aspects, such as the level at which the routing decisions are made. In our case, customers (SPs) access the secondary facilities (CCs) by themselves, so vehicle routing decisions are not relevant at this level. We determine the routes that originate from the primary (biodiesel) facility, visit the secondary facilities (CCs), and end at the biodiesel facility. Moreover, bin allocation decisions bring additional challenges to our problem. Specifically, the WCO is accumulated in the bins, and each bin has a fixed unit cost. Hence, rather than just satisfying the facility capacity constraints, we also try to use the minimum number of bins while assigning SPs to the CCs to be opened. Moreover, vehicles have capacity constraints in terms of bins, so to use a lower number of vehicles and decrease transportation costs, bin allocation decisions must be made carefully. We present a mathematical model and a decomposition-based heuristic that integrate the location, assignment, routing, and bin allocation decisions effectively.

The remainder of this paper is organized as follows. In Sect. 6.2, we review the relevant literature. In Sect. 6.3, we describe our problem and present the mathematical model. We propose our decomposition-based heuristic in Sect. 6.4 and present the numerical results in Sect. 6.5. Finally, we conclude our work in Sect. 6.6.

## 6.2   Literature Review

In this section, we review the relevant literature in oil collection problems, the broader waste collection problems, and the LRP.

There exist several studies that particularly focus on waste oil collection operations. Janssens [12] presents a mathematical model to determine the fleet size in a waste oil collection environment. Two models are developed to estimate (1) the amount of waste oil to be collected and (2) the number of routes per year to the communities and the time required to travel to the waste oil producers to fill and empty the truck. Ramos et al. [13] study a WCO collection system where multiple depots are used and open routes between depots are allowed. A mixed-integer linear programming (MILP) is presented where capacity and time duration constraints are also considered. Aksen et al. [14] study a selective and periodic inventory routing problem motivated by a biodiesel company in Istanbul that utilizes the collected amount of waste vegetable oil as raw material for production. The aim of this problem is to minimize the total collection, inventory, and purchasing cost while informing the company about which source points to be included in the collection program, which of them to be visited on each day, which periodic route planning to be executed repeatedly over an infinite horizon, and number of vehicles to operate. Aksen et al. [15] develop an adaptive large neighborhood search algorithm for the former problem. Zhao and Verter [16] use a modified weighted goal programming approach to solve a bi-objective LRP aiming to minimize the total environmental risk and the total cost in a used oil collection and processing system. Besides the location and routing decisions concerning both the storage and integrated facilities, optimal capacity levels are determined as well. Weyland et al. [17] study a variant of a periodic vehicle routing problem (PVRP) focusing on used oil collection through hotels and restaurants and solve it by a local search algorithm. Zhang and Jiang [18] deal with a three-level network consisting of WCO suppliers, integrated bio-refineries and demand zones. Biodiesel price is deemed uncertain, thus robust optimization methods are utilized to solve this multi-objective MILP. Waste management supply chain includes collecting, recycling, and/or recovery efforts which create a reverse material flow, thus it can be considered as reverse logistics problems [19]. Fonseca et al. [20] present a simulation study in which Geographic Information System (GIS) tools are used to determine routes for WCO collection. Based on the simulation results, it is seen that instead of suppliers calling and asking for the company to pick up the oil, the company should use a Milk Run model, in which they visit multiple suppliers and collect the oil containers, even when they are not completely full. Repoussis et al. [21] develop a web-based decision support system (DSS) for a waste lube oils collection and recycling case in Greece. In this study, vehicles are compartmentalized in order to avoid different quality oils getting blended. A list-based threshold accepting metaheuristic is proposed as a solution approach. Supsomboon [22] develops an optimization tool using Visual Basic Interface for solving VRP for used oil collection. Different than these studies, we focus on developing an effective system for collecting WCO from households and address bin allocation decisions along with location and routing decisions. We are particularly motivated by the WCO collection network and the problem described in [10], and present an efficient and practical solution method to design an effective WCO collection network.

Although oil collection literature is rather limited, waste collection literature is rich and provides a vast variety of problems. The waste collection literature covers the studies that focus on the collection of municipal solid waste, industrial waste, and commercial waste. Beliën et al. [23] and Han and Cueto [24] present reviews for the regarding literature with a particular focus on the vehicle routing problems. Hemmelmayr et al. [25] study a Waste Bin Allocation Routing Problem (WBARP) where site locations are known a priori. The WBARP focuses on two main decisions. The first focus is on routing where service choices (SC) are incorporated and intermediate facilities (IF) for dumping are almost always used, thus this part of the problem is named PVRP-IF-SC. The second focus is on determining the number of bins to place at each collection site, given the restrictions on space and the total number of bins available. Two solution approaches are combined to solve this problem; an effective variable neighborhood search metaheuristic for the routing part and a mixed-integer linear programming based exact method for the bin allocation part. Angelelli and Speranza [26] also study a PVRP-IF in which a vehicle's capacity is renewed whenever it visits an intermediate facility. Teixeira et al. [27] propose a three-phase heuristic algorithm for a PVRP for the separate collection of three types of waste: glass, paper, and plastic/metal. In this approach, while sequencing decisions are made, delivery zones and waste types to be collected are already determined respectively for each day. Buhrkala et al. [28], Benjamin and Beasley [29], and Kim et al. [30] propose different heuristic algorithms for the waste collection vehicle routing problem with time windows where driver rest periods are taken into consideration. Mes et al. [31] consider the problem of dynamic waste collection through sensor-equipped underground containers. Their problem is a well-known reverse IRP possessing the routing and container selection decisions. Miranda et al. [32] study a bi-objective insular traveling salesman problem where a set of rural islands needs to be visited for household collection purposes and the maritime transportation and ground transportation costs are minimized. Ramos et al. [33] introduce a new objective function of minimizing $CO_2$ emissions. They conclude that the total distance traveled and $CO_2$ emissions are highly correlated, thus minimizing the total distance with economic concerns is concurrently for the benefit of mitigating the environmental damage of transportation. In waste collection literature, similarly to our problem, it is very common to use intermediate facilities for dumping or collection purposes. Even if our problem setting shows similarities with the existent LRP studies, incorporation of the bin allocation decisions and the echelon at which routing decisions are made differentiate our problem.

The class of problems that our study falls into is the LRP, which has been studied for almost 50 years. A vast amount of solution approaches have been proposed for this problem, which are reviewed by Prodhon and Prins [34]. We review some related LRP studies that use similar decomposition-based solution methods. Wu et al. [35] solve a multi-depot LRP by dividing the problem into location-allocation and general vehicle routing subproblems. Each subproblem is then sequentially and iteratively solved by the embedded simulated annealing algorithm. Perl and Daskin [36] develop a heuristic algorithm for the warehouse location-routing problem by decomposing it into three subproblems and sequentially solving them either to optimality or heuristically while observing the dependence between them. The first phase is a multi-depot vehicle-dispatch problem, which is followed by a location-allocation problem and then a multi-depot routing-allocation problem is solved.

Two-echelon LRP (2E-LRP) models arise in the reverse logistics literature when location decisions are made to determine the first and the second echelon facilities [37–39]. In such problems, routes are constructed at two levels. The primary routes are for transporting items between the facilities at the first and the second levels, whereas the secondary routes are for serving customers from the second-level facilities [40]. Vidovic et al. [41] study a 2E-LRP model in the context of non-hazardous recyclables collection, where the locations of transfer stations and collection points are determined in the first and second echelons, respectively. They aim to maximize the total profit obtained by the collected recyclables.

The most similar variant of LRP to our problem is the widely studied school bus routing problem (SBRP), which deals with finding the optimal schedule for a fleet of school buses that transport students in between their designated bus stop and school [42]. As such, in our problem, customers (students) arrive in their designated secondary facility with their own means and routing decisions occur in between the first facility (school) and the secondary facilities (bus stops). In the SBRP, students are assigned to the bus stops and they get on the buses. On the other hand, in our problem SPs bring their WCO in liters. We assign SPs to the CCs so that the accumulated WCO is placed into the minimum number of bins, which are then loaded into vehicles. Thus, our problem becomes more complicated since we try to better utilize the bins as an extra decision while assigning the supply nodes. This characteristic distinguishes our problem and requires developing new solution methods. Ellegood et al. [43] recently present a literature review where 64 new SBRP research publications are analyzed and classified according to the sub-problem type, problem characteristics (constraints, objective function), and solution approaches.

The main contribution of this study is that we present a new variant of secondary-facility LRP, which covers the decisions that have not previously been considered together. Mainly, the incorporation of the bin allocation decision makes our problem more complex than the secondary-facility LRP. We next present a mathematical model and a solution method for this new problem.

## 6.3   Problem Description and Mathematical Model

Our study is motivated by a real-world WCO collection system, in which households bring their WCO to the CCs, which are located at easily accessible public facilities such as schools, mosques, municipality buildings, etc., and deposit the oil into the bins [10]. We group the households based on their regions so that each SP represents a group of households. The amount of WCO produced by each SP is assumed to be known, which can be estimated from historical data. There is a limit on the number of bins that each CC can accommodate since there is a certain space allocated for this purpose. Since people would not prefer traveling long distances for recycling activities and assigning people to remote centers may decrease the total amount of oil collected, in our network each SP must be assigned to one CC within a reasonable walking distance. The oil collection is performed periodically. Specifically, on the last day of each time period (week), a homogeneous fleet of capacitated vehicles visits the CCs and replaces the filled bins with the empty ones regardless of how much oil accumulated in the bins. Each CC can only be visited by one vehicle during the collection day, that is split delivery is not allowed. Vehicles start and end their routes at the biodiesel facility where the collected WCO is used as raw material for production. Since truck drivers cannot work beyond their working hours, each vehicle can travel for the specified working time. Multiplying this time with the average truck speed will yield a restriction on the total travel distance for each vehicle. There is a cost incurred anytime a CC is opened and a bin is placed at a CC. We seek for the least costly solution while determining the CCs to be opened, the number of bins to allocate at each CC, the assignment of each SP to one of the CCs and the vehicle routes.

Considering bin allocation decisions and minimizing costs associated with the number of bins used prevent SPs from being randomly assigned to a CC within the walking distance threshold. In Fig. 6.2, we present a simple example to illustrate the effect of bin allocation decisions and costs on WCO network design decisions. Suppose that there are two CCs with a capacity of two bins for each and five SPs which can reach to both CCs. SPs have a WCO supply of 10, 42, 30, 20, and 10 liters, respectively. We assume that each bin has a capacity of 50 liters and the cost of placing each bin is the same. If minimizing the total bin cost was not in question, assigning the first two SPs to one CC and the remainders to the other CC would be a feasible assignment (Fig. 6.2a). In that case four bins would be used in total. However, the optimal bin allocation would be obtained by assigning SP-1, SP-3, and SP-5 to one CC and SP-2 and SP-4 to the other one, which results in three bins to be used in total (Fig. 6.2b). In this case, by optimizing the bin allocation decision we reduce the total bin cost. Moreover, since the number of bins that each vehicle can carry is limited, using less number of bins may help decrease routing costs. Likewise, since each CC has a capacity in terms of bins, a wise assignment of SPs might affect the location decisions and help achieve a better facility opening cost.

We formulate a mixed-integer mathematical model for our problem. We let $P$ denote the number of candidate CCs and define the set of CCs as $\mathcal{P} = \{1, \ldots, P\}$.

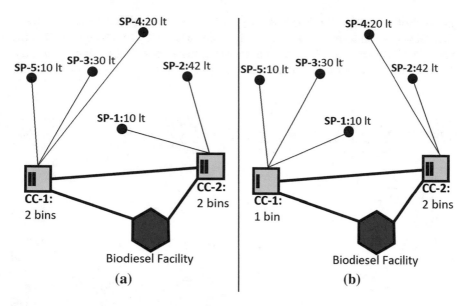

**Fig. 6.2** An illustration of bin allocation

We represent the biodiesel facility within the $\mathcal{P}_0$ set which is $\mathcal{P}_0 = \{0, \ldots, P\}$. We let $S$ denote the number of SPs and define the set of SPs as $\mathcal{S} = \{1, \ldots, S\}$. We let $V$ denote the size of the vehicle fleet and define the set of vehicles as $\mathcal{V} = \{1, \ldots, V\}$. We let $\mathcal{N}_{P_i}$ to denote the set of SPs that are within the walking distance of CC $i$. We let $\mathcal{N}_{S_j}$ to denote the set of CCs that SP $j$ can reach to. We use the following decision variables in our formulation:

$$x_i = \begin{cases} 1, & \textit{if CC } i \in \mathcal{P} \textit{ is opened,} \\ 0, & \textit{otherwise,} \end{cases}$$

$y_i$ = Number of bins placed at CC $i \in \mathcal{P}$,

$$z_{ij} = \begin{cases} 1, & \textit{if SP } j \in \mathcal{S} \textit{ is assigned to CC } i \in \mathcal{P}, \\ 0, & \textit{otherwise,} \end{cases}$$

$$v_{ijk} = \begin{cases} 1, & \textit{if vehicle } k \in \mathcal{V} \textit{ goes from node } i \in \mathcal{P}_0 \textit{ to node } j \in \mathcal{P}_0 \\ 0, & \textit{otherwise,} \end{cases}$$

$t_{ik}$ = Number of bins placed at CC $i \in \mathcal{P}$, if it is visited by vehicle $k \in \mathcal{V}$

$u_i$ = Auxiliary variable

The parameters are introduced next. Supply of SP $j$ is given by $d_j$. The travel distance between nodes is represented by $c_{ij}$. Each vehicle is allowed to travel at most $R$ distance units. $\alpha$ is the traveling cost per distance unit. We use $f_i$ to denote the cost of opening $CC$ $i$. The cost of placing a bin to a CC is represented by $b$. The maximum number of bins that can be located at a CC is given by $C_i$. $B$ is the capacity of a bin in liters. $H$ is the maximum number of bins that a vehicle can carry. $M$ refers to the largest possible number that $t_{ik}$ can take, which in this case equals to $H$.

$$\min \quad \alpha \sum_{i \in \mathcal{P}_0} \sum_{j \in \mathcal{P}_0} \sum_{k \in \mathcal{V}} v_{ijk} c_{ij} + \sum_{i \in \mathcal{P}} x_i f_i + b \sum_{i \in \mathcal{P}} y_i \tag{6.1}$$

$$st. \quad \sum_{j \in N_{P_i}} z_{ij} d_j \le B y_i \qquad\qquad \forall i \in \mathcal{P} \tag{6.2}$$

$$y_i \le C_i x_i \qquad\qquad \forall i \in \mathcal{P} \tag{6.3}$$

$$z_{ij} - x_i \le 0 \qquad\qquad \forall i \in \mathcal{P}, j \in N_{P_i} \tag{6.4}$$

$$z_{ij} = 0 \qquad\qquad \forall i \in \mathcal{P}, j \in \{\mathcal{S} \backslash N_{P_i}\} \tag{6.5}$$

$$\sum_{i \in N_{S_j}} z_{ij} = 1 \qquad\qquad \forall j \in \mathcal{S} \tag{6.6}$$

$$\sum_{i \in \mathcal{P}} v_{0ik} \le 1 \qquad\qquad \forall k \in \mathcal{V} \tag{6.7}$$

$$t_{ik} \le M \sum_{j \in \mathcal{P}_0} v_{jik} \qquad\qquad \forall i \in \mathcal{P} : i \ne j, k \in \mathcal{V} \tag{6.8}$$

$$t_{ik} \le y_i \qquad\qquad \forall i \in \mathcal{P}, k \in \mathcal{V} \tag{6.9}$$

$$\sum_{i \in \mathcal{P}} t_{ik} \le H \qquad\qquad \forall k \in \mathcal{V} \tag{6.10}$$

$$\sum_{j \in \mathcal{P}_0} v_{jik} = \sum_{l \in \mathcal{P}_0} v_{ilk} \qquad\qquad \forall i \in \mathcal{P}_0, k \in \mathcal{V} \tag{6.11}$$

$$\sum_{j\in\mathcal{P}_0}\sum_{k\in\mathcal{V}} v_{ijk} = x_i \qquad\qquad \forall i \in \mathcal{P} : i \neq j$$

$$(6.12)$$

$$\sum_{j\in\mathcal{P}_0}\sum_{k\in\mathcal{V}} v_{jik} = x_i \qquad\qquad \forall i \in \mathcal{P} : i \neq j$$

$$(6.13)$$

$$u_i - u_j + P \sum_{k\in\mathcal{V}} v_{ijk} \leq P - 1 \qquad\qquad \forall i, j \in \mathcal{P} : i \neq j$$

$$(6.14)$$

$$\sum_{i\in\mathcal{P}_0}\sum_{j\in\mathcal{P}_0} v_{ijk} c_{ij} \leq R \qquad\qquad \forall k \in \mathcal{V}$$

$$(6.15)$$

$$x_i, z_{ij} \in \{0, 1\} \qquad\qquad \forall i \in \mathcal{P}, j \in \mathcal{S}$$
$$(6.16)$$

$$v_{ijk} \in \{0, 1\} \qquad\qquad \forall i \in \mathcal{P}_0, j \in \mathcal{P}_0, k \in \mathcal{V}$$
$$(6.17)$$

$$y_i, t_{ik} \geq 0, integer \qquad\qquad \forall i \in \mathcal{P}, k \in \mathcal{V}$$
$$(6.18)$$

$$u_i \geq 0 \qquad\qquad \forall i \in \mathcal{P}$$
$$(6.19)$$

Equation (6.1) is the objective function, which minimizes the total cost including the routing cost, fixed cost of opening CCs, and bin allocation cost. Constraints (6.2) ensure that the capacity of each CC is not exceeded by the total amount of WCO brought by the assigned SPs. Constraints (6.3) set an upper bound on the maximum number of bins to be allocated to each CC. Constraints (6.4) prevent an SP from being assigned to a closed CC. Constraints (6.5) and (6.6) make sure that each SP is assigned to a CC within the walking distance. Constraints (6.7) ensure that each vehicle can only make one tour per period. Constraints (6.8) and (6.9) relate to $t_{ik}$ integer variable so that if a vehicle does not visit a CC or there are no bins allocated to a CC, then $t_{ik}$ must be zero. Constraints (6.10) make sure that the vehicle capacity is respected for each route. Constraints (6.11) are route continuity constraints. Constraints (6.12) and (6.13) ensure that a vehicle is routed through a CC if and only if that CC is opened. Constraints (6.14) are the MTZ subtour elimination constraints [44]. Constraints (6.15) guarantee that no route can be longer than the maximum route length. Constraints (6.16), (6.17), (6.18), and (6.19) are the non-negativity and integrality constraints.

Since our problem is a variant of the LRP, which is NP-Hard, solving the proposed mathematical model optimally for large instances would not be possible. Therefore, we resort to developing an efficient heuristic algorithm to solve our problem, described in the next section.

## 6.4   Solution Approach

Our problem is a variant of the secondary-facility LRP. Location-routing problems are composed of facility location and vehicle routing problems which are both categorized under the class of NP-hard problems [45]. Thus, LRP is known to be NP-hard as well. Together with the newly introduced challenges of our problem, the inherent complexity of LRP requires effective solution approaches. In this regard, we favor the decomposition technique which enables us to handle the problem by dividing it into two subproblems. Location, assignment, and bin allocation decisions are addressed in the first subproblem, whose solution is fed into the second subproblem that focuses on routing decisions. Hereby different aspects considered in a problem can be solved independently and interact with each other. In the decomposition algorithm presented in this study, we solve the first subproblem multiple times to obtain different sets of CCs. Each time the first model is rerun, the previously obtained sets are prevented from showing up in the following runs. The resulting set of CCs is then provided to the VRP model and the one that generates the least transportation cost is agreed to be the best solution found by this approach. Different solutions are evaluated in this manner until the total solution time limit is reached.

The first subproblem involves constraints (6.2)–(6.6) in the base model, and the integrality and non-negativity constraints for the existing decision variables which are $x_i$, $z_{ij}$, and $y_i$ are respected. The objective in the first subproblem is to find the set of CCs that generate the least costly solution in terms of the CC opening cost and the bin cost. Therefore the transportation cost component in the base model is excluded for this subproblem. Two extra constraints are added to the model of the first subproblem. The first one (6.20) is included to obtain a bin allocation that does not surpass the maximum number of bins that the fleet can carry. The necessity of this restriction arises from the fact that the first subproblem does not consider any routing decisions; therefore, infeasible assignments might occur such that fleet capacity is exceeded.

$$\sum_{i \in \mathcal{P}} y_i \leq VH \tag{6.20}$$

Constraints (6.21) are added to allow generating different solutions and achieve a better objective value. Specifically, constraints (6.21) prevent the model from coming up with the same sets of CCs as the previous ones. Let $w \in \mathcal{W}$ be the sets of CCs that are obtained in each run within the given time limit. $\theta_w$ denotes the length of each set $w$, in other words, the number of CCs that are opened in set $w$. Constraints (6.21) are used starting from the second run of the model (i.e., after an initial solution is obtained).

$$\sum_{i \in w} x_i \leq \theta_w - 1 \qquad\qquad \forall w \in \mathcal{W} \qquad\qquad (6.21)$$

The second subproblem is a generic vehicle routing problem. We adapt the three index formulation proposed by Golden [46] where maximum route duration restriction is respected. The steps of DA are represented in Algorithm 1.

---

**Algorithm 1:** Decomposition algorithm

---

1   Initialize $\mathcal{W}$;
2   **while** *time limit is not reached* **do**
3       Give $\mathcal{W}$ to the assignment model;
4       //sets of CCs in $\mathcal{W}$ are prevented from opening
5       Solve the assignment model;
6       **if** *the assignment is feasible* **then**
7           Obtain the set of opened CCs from the solution as $w$;
8           Add $w$ to $\mathcal{W}$;
9           Solve the VRP model with obtained solution;
10          **if** *the VRP solution is feasible* **then**
11              Keep the result;
12      **else**
13          *break*;

14  Report the best solution among the stored results;

---

## 6.5   Computational Study

In this section, we compare the performance of the proposed algorithm (DA) with the proposed mixed-integer programming model (MIP) and discuss the results of the computational experiments. We develop a set of instances to test our approach. Specifically, we generate 20 random instances over a $1000 \times 1000$ square. The first half of the instances has 20 CCs and 100 SPs, while the other half has 40 CCs and 200 SPs. We set the truck capacity to 24 bins and bin capacity to 50 l. We assume the size of the fleet is 10 and the cost of a bin is 10. Based on the solutions obtained during preliminary experiments, we set the maximum walking distance as 300 and maximum driving distance as 1800. We generate clusters for each instance to represent the dense regions. Instances with 20 CCs have 10 clusters, while others have 20 clusters. Clusters are distributed uniformly throughout the map. The cluster ratio controls how many of the total nodes including all SPs and CCs fall within a cluster. We set the ratio as 0.5 meaning that 50% of all the nodes fall into a cluster.

The position of each node in a cluster is generated by using a standard normal distribution. The remaining nodes are distributed uniformly throughout the map.

We conduct a computational study by solving each of the 20 problem instances with the proposed solution methods. We solve MIP with CPLEX solver with a time limit of 1 h. We conduct all computational experiments on a system that has a 1.80 GHz i7-8550U processor and an 8GB RAM. We implement both of the algorithms in Python and use CPLEX 12.8 as a solver with default settings. The entire algorithm has a time limit of 10 min. For the first subproblem of the decomposition algorithm, we set a 3% optimality gap limit to increase the number of solutions obtained within the specified time restriction. Since routing cost dominates the other cost components and is highly effective on the objective value, no optimality gap is set for the second subproblem. Thus, each VRP instance is solved to optimality unless the time limit is reached.

In Table 6.1, we present the results of the proposed solution methods. The first three columns represent the identification (ID), the number of candidate CCs ($P$), and the number of candidate SPs ($S$) for every instance. The column "T" represents the runtime of the MIP model in seconds, while the column "G" represents the optimality gap achieved by CPLEX in solving the mathematical model. The table also reports the objective values of the model (column "MIP") and the proposed decomposition algorithm (column "DA"). The best solution found for that instance

**Table 6.1** Comparison of MIP and DA

| ID | P | S | T(sec) | G(%) | MIP | DA | NB | DF(%) |
|----|----|-----|--------|-------|---------|---------|-----|-------|
| 1 | 20 | 100 | 73.17 | 0 | **2770.47** | 3028.62 | 107 | −8.52 |
| 2 | 20 | 100 | 3600 | 8.28 | **3600.78** | **3600.78** | 84 | 0 |
| 3 | 20 | 100 | 414.75 | 0 | **3381.08** | **3381.08** | 86 | 0 |
| 4 | 20 | 100 | 36.52 | 0 | **2997.77** | **2997.77** | 117 | 0 |
| 5 | 20 | 100 | 3600 | 34.15 | 5230.02 | **5221.09** | 35 | 0.17 |
| 6 | 20 | 100 | 238.55 | 0 | **2947.29** | **2947.29** | 100 | 0 |
| 7 | 20 | 100 | 20.95 | 0 | **2329.80** | **2329.80** | 84 | 0 |
| 8 | 20 | 100 | 119.42 | 0 | **2886.91** | **2886.91** | 129 | 0 |
| 9 | 20 | 100 | 1661 | 0 | **3679.52** | **3679.52** | 82 | 0 |
| 10 | 20 | 100 | 1755.5 | 0 | **3315.70** | **3315.70** | 92 | 0 |
| 11 | 40 | 200 | 3600 | 45.98 | 6211.79 | **5084.84** | 9 | 22.16 |
| 12 | 40 | 200 | 3600 | 36.81 | 6158.61 | **4967.89** | 8 | 23.97 |
| 13 | 40 | 200 | 3600 | 41.52 | 5470.45 | **5306.80** | 8 | 3.08 |
| 14 | 40 | 200 | 3600 | 33.34 | 4963.19 | **4463.94** | 10 | 11.18 |
| 15 | 40 | 200 | 3600 | – | – | **5436.55** | 8 | – |
| 16 | 40 | 200 | 3600 | 36.36 | 5476.12 | **4781.90** | 9 | 14.52 |
| 17 | 40 | 200 | 3600 | – | – | **8314.98** | 8 | – |
| 18 | 40 | 200 | 3600 | 31.93 | 4776.59 | **4665.02** | 8 | 2.39 |
| 19 | 40 | 200 | 3600 | 42.01 | 5529.73 | **5077.65** | 8 | 8.9 |
| 20 | 40 | 200 | 3600 | – | – | **4858.71** | 8 | – |

among the two methods is marked in bold. The column "NB" represents the number of solutions generated for DA within the time limit. Lastly, the column "DF" represents the difference between objective values of MIP and DA in percentage. DA stops if either all the feasible assignments are evaluated within the given time or the time limit is reached. For all the instances, we observe that DA stops due to the time limit, which is 10 min.

We observe that the size of the instance has a huge impact on the solution quality of the MIP. The average optimality gap is 4.2% for the first 10 instances. For the larger instances, the average optimality gap is 37.8%. A feasible solution could not be found in three of the large instances within 1-h solution time limit. When we compare the solutions obtained by MIP and DA, we see that DA finds the same or better solution compared to MIP in 95% of the instances. This difference appears the most in large instances (no. 11–20) since DA finds a better solution than MIP in every one of them which constitutes 50% of all the test instances. These results suggest that DA tends to perform better than MIP, especially for larger instances. The column "NB" in Table 6.1 gives us some insight about the impact of instance size on the DA algorithm. The average number of solutions generated in 10 min is 92 for the small-sized instances (no. 1–10), while it is 8 for the large-sized instances (no. 11–20). Doubling the instance size causes such a significant decline in the number of solutions found, yet DA performs better than the proposed model.

Two of the small-sized instances (no. 1 and 5) generate different objective values. In Table 6.2, each of the cost components for both instances is reported. "$\Delta$" column represents the difference between MIP and DA. To make it clear, the first subproblem of DA aims to minimize the total fixed and bin cost while making the location, assignment, and bin allocation decisions whereas no routing decisions are considered. Hence in both instances, DA generates a better fixed cost. In instance #1 the reason why MIP performs better than DA is that the $\Delta$ in the routing cost is higher than the $\Delta$ in the total fixed cost. On the other hand, in instance #5 the $\Delta$ in the total fixed cost is large enough to compensate the $\Delta$ in the routing cost.

As indicated by these results, the proposed DA can achieve significantly superior solutions than the optimization solver. Moreover, the solution times of the DA are favorable. There exists only one instance where the DA could not find a better solution than the benchmark. However, the solutions may be improved by extending the time limit of the DA algorithm to generate a large number of solutions in the first subproblem.

**Table 6.2** Cost components of instances 1 and 5

|  | Instance #1 | | | Instance #5 | | |
|---|---|---|---|---|---|---|
|  | MIP | DA | $\Delta$ | MIP | DA | $\Delta$ |
| Fixed cost | 829 | 652 | 177 | 655 | 637 | 18 |
| Routing cost | 1,731.5 | 2,166.6 | −435.1 | 4,355.0 | 4,364.1 | −9.1 |
| Bin cost | 210 | 210 | 0 | 220 | 220 | 0 |

## 6.6   Conclusion

This study focuses on designing a WCO collection network to collect the WCO accumulated at the household level. We are motivated by the oil collection operations of a biodiesel company, which uses the collected WCO as a raw material in biodiesel production. In this way, while households can dispose their used cooking oil without damaging the environment or the infrastructure, biodiesel companies meet their raw material needs in a much cheaper way than using virgin oil. The problem addressed in this study is a new variant of the secondary-facility location-routing problem where the bin allocation decision is also under consideration. Bin placement is subjected to a cost per unit which requires to aggregate the collected WCO into lower number of bins, hence the problem becomes more challenging. We propose a MIP model which addresses the restrictions inspired by the real-world applications. Due to the complex structure of our problem, we develop a decomposition-based heuristic (DA) approach, which divides the problem into two subproblems. We evaluate the results of MIP and DA with respect to both solution quality and computational time. Our numerical results show that DA can provide better solutions than MIP, especially for large-sized instances. DA finds the same solution with MIP in 40% of the test instances while it performs better than MIP in 55% of the test instances created. Since oil collection operations at the household level has not received much attention in the literature, we hope that our study will lead to future studies that investigate this problem from both practical and theoretical perspectives. For example, future research can focus on developing other solution approaches for the proposed problem. Moreover, in our study, we assume that the SPs will travel to their assigned CCs. Future research can explore behavioral aspects related to WCO recycling for the households and incorporate the relevant aspects in decision making. Furthermore, simple local search moves can be designed and used to improve the DA solutions.

## References

1. Phan, A.N., Phan, T.M.: Biodiesel production from waste cooking oils. Fuel **87**(17–18), 3490–3496 (2008)
2. Lean, G.: Oil and gas may run short by 2015. The Independent, UK. http://environment.independent.co.uk/climate_change/article2790960.ece (Accessed on 23 July 2007)
3. Demirbas, A.: Biodegradability of Biodiesel and Petrodiesel Fuels. Energy Sources Part A: Recovery Utilization Environmental Effects **31**(2), 169–174 (2008)
4. Chhetri, A.B., Watts, K.C., Islam, M.R.: Waste cooking oil as an alternate feedstock for biodiesel production. Energies **1**(1), 3–18 (2008)
5. Nas, B., Berktay, A.: Energy potential of biodiesel generated from waste cooking oil: An environmental approach. Energy Sources Part B: Economics Planning Policy **2**(1), 63–71 (2007)
6. Gonzalez-Gomez, M.E., Howard-Hildige, R., Leahy, J.J., Reilly, T.O'., Supple, B., Malone, M.: Emission and performance characteristics of a 2 litre Toyota diesel van operating on esterified waste cooking oil and mineral diesel fuel. Environ. Monit. Assess. **65**(1–2), 13–20 (2000)

7. Peiro, L.T., Lombardi, L., Mendez, G.V., Durany, X.G.: Life cycle assessment (LCA) and exergetic life cycle assessment (ELCA) of the production of biodiesel from used cooking oil (UCO). Energy **35**(2), 889–893 (2010)
8. Ma, F., Hanna, M.A.: Biodiesel production: a review. Bioresource Technology **70**(1), 1–15 (1999)
9. Zhang, Y., Dube, M.A., McLean, D.D., Kates, M.: Biodiesel production from waste cooking oil: 2. Economic assessment and sensitivity analysis. Bioresource Technology **90**(3), 229–240 (2003)
10. Karabak, F.: A Location-Routing Problem for Waste Oil Collection. MS Thesis, Ozyegin University (2016)
11. Min, H., Jayaraman, V., Srivastava, R.: Combined location-routing problems: A synthesis and future research directions. Eur. J. Oper. Res. **108**(1), 1–15 (1998)
12. Janssens, G.K.: Fleet size determination for waste oil collection in the province of Antwerp. Yugoslav J. Oper. Res. **3**(1), 103–113 (1993)
13. Ramos, T.R.P., Gomes, M.I., Barbosa-Póvoa, A.P.: Planning waste cooking oil collection systems. Waste Management **33**(8), 1691–1703 (2013)
14. Aksen, D., Kaya, O., Salman, F.S., Akca, Y.: Selective and periodic inventory routing problem for waste vegetable oil collection. Optimization Letters **6**(6), 1063–1080 (2012)
15. Aksen, D., Kaya, O., Salman, F.S., Tuncel, O.: An adaptive large neighborhood search algorithm for a selective and periodic inventory routing problem. Eur. J. Oper. Res. **239**(2), 413–426 (2014)
16. Zhao, J., Verter, V.: A bi-objective model for the used oil location-routing problem. Comput. Oper. Res. **62**, 157–168 (2015)
17. Weyland, D., Salani, M., Montemanni, R., Gambardella, L.M.: Vehicle routing for exhausted oil collection. J. Traffic Logist. Eng. **1**(1), 5–8 (2013)
18. Zhang, Y., Jiang, Y.: Robust optimization on sustainable biodiesel supply chain produced from waste cooking oil under price uncertainty. Waste Management **60**, 329–339 (2016)
19. Zhang, Y.M., Huang, G.H., He, L.: An inexact reverse logistics model for municipal solid waste management systems. J. Environ. Manag. **92**(3), 522–530 (2011)
20. Fonseca, A.G., Oliveira, R.L., Lima, R.S.: Structuring reverse logistics for waste cooking oil with geographic information systems. Proc. CUPUM (2013)
21. Repoussis, P.P., Paraskevopoulos, D.C., Zobolas, G., Tarantilis, C.D., Ioannou, G.: A web-based decision support system for waste lube oils collection and recycling. Eur. J. Oper. Res. **195**(3), 676–700 (2009)
22. Supsomboon, S.: A mathematical model for vehicle routing of used-oil collection in bio-diesel production using visual basic interface: village bank and bio-diesel project. KKU Eng. J. **37**(2), 151–159 (2010)
23. Beliën, J., De Boeck, L., Van Ackere, J.: Municipal solid waste collection and management problems: A literature review. Transportation Science **48**(1), 78–102 (2012)
24. Han, H., Ponce-Cueto, E.: Waste collection vehicle routing problem: literature review. Promet-Traffic Transp. **27**(4), 345–358 (2015)
25. Hemmelmayr, V.C., Doerner, K.F., Hartl, R.F., Vigo, D.: Models and algorithms for the integrated planning of bin allocation and vehicle routing in solid waste management. Transportation Science **48**(1), 103–120 (2014)
26. Angelelli, E., Speranza, M.G.: The periodic vehicle routing problem with intermediate facilities. Eur. J. Oper. Res. **137**(2), 233–247 (2002)
27. Teixeira, J., Antunes, A.P., de Sousa, J.P.: Recyclable waste collection planning-a case study. Eur. J. Oper. Res. **158**(3), 543–554 (2004)
28. Buhrkala, K., Larsena, A., Ropke, S.: The waste collection vehicle routing problem with time windows in a city logistics context. Procedia Soc. Behav. Sci. **39**, 241–254 (2012)
29. Benjamin, A.M., Beasley, J.E.: Metaheuristics for the waste collection vehicle routing problem with time windows, driver rest period and multiple disposal facilities. Comput. Oper. Res. **37**(12), 2270–2280 (2010)

30. Kim, B-I., Kim, S., Sahoo, S.: Waste collection vehicle routing problem with time windows. Comput. Oper. Res. **33**(12), 3624–3642 (2006)
31. Mes, M., Schutten, M., Rivera, A.P.: Inventory routing for dynamic waste collection. Waste Management **34**(9), 1564–1576 (2014)
32. Miranda, P.A., Blazquez, C.A., Obreque, C., Maturana-Ross, J., Gutierrez-Jarpa, G.: The bi-objective insular traveling salesman problem with maritime and ground transportation costs. Eur. J. Oper. Res. **271**(3), 1014–1036 (2018)
33. Ramos, T.R.P., Gomes, M.I., Barbosa-Póvoa, A.P.: Economic and environmental concerns in planning recyclable waste collection systems. Transp. Res. Part E **62**, 34–54 (2014)
34. Prodhon, C., Prins, C.: A survey of recent research on location-routing problems. Eur. J. Oper. Res. **238**(1), 1–17 (2014)
35. Wu, T.H., Low, C., Bai, J.W.: Heuristic solutions to multi-depot location-routing problems. Comput. Oper. Res. **29**(10), 1393–1415 (2002)
36. Perl, J., Daskin, M.S.: A warehouse location-routing problem. Transp. Res. Part B: Methodol. **19**(5), 381–396 (1985)
37. Govindan, K., Jafarian, A., Khodaverdi, R., Devika, K.: Two-echelon multiple-vehicle location-routing problem with time windows for optimization of sustainable supply chain network of perishable food. Int. J. Prod. Econ. **152**, 9–28 (2014)
38. Dondo, R.G., Mendez, C.A.: Operational planning of forward and reverse logistic activities on multi-echelon supply-chain networks. Comput. Chem. Eng. **88**, 170–184 (2016)
39. Ghezavati, V.R., Beigi, M.: Solving a bi-objective mathematical model for location-routing problem with time windows in multi-echelon reverse logistics using metaheuristic procedure. J. Ind. Eng. Int. **12**(4), 469–483 (2016)
40. Rahmani, Y., Cherif-Khettaf, W.R., Oulamara, A.: The two-echelon multi-products location-routing problem with pickup and delivery: formulation and heuristic approaches. Int. J. Prod. Res. **54**(4), 999–1019 (2016)
41. Vidovic, M., Ratkovic, B., Bjelic, N., Popovic, D.: A two-echelon location-routing model for designing recycling logistics networks with profit: MILP and heuristic approach. Expert Syst. Appl. **51**, 34–48 (2016)
42. Park, J., Kim, B.-I.: The school bus routing problem: A review. Eur. J. Oper. Res. **202**(2), 311–319 (2010)
43. Ellegood, W.A., Solomon, S., North, J., Campbell, J.F.: School bus routing problem:Contemporary trends and research directions. Omega (2019, In Press)
44. Miller, C.E., Tucker, A.W., Zemlin, R.A.: Integer programming formulations and traveling salesman problems. J. Assoc. Comput. Mach. **7**(4), 326–329 (1960)
45. Karp, R.M.: Reducibility among combinatorial problems. In: Miller, R.E., Thatcher, J.W., Bohlinger, J.D. (eds.) Complexity of Computer Computations, pp. 85–103. Plenum Press (1972)
46. Golden, B.L.: Vehicle routing problems: formulations and heuristic solution techniques. M.I.T. Operations Research Center Technical Report, vol. 113 (1975)

# Chapter 7
# Time-Dependent Green Vehicle Routing Problem

Golnush Masghati-Amoli and Ali Haghani

**Abstract** This chapter describes a model for a time-dependent green vehicle routing problem (TDGVRP) with a mixed fleet of electric (ECV) and internal combustion engine commercial vehicles (ICCV) that finds the optimal fleet design and routes for last-mile delivery operations. Due to the high sensitivity of electric vehicle's driving range to energy requirements, the effect of congestion on vehicle energy consumption is accounted for by time dependency of travel time. The GVRP model proposed in this study takes into account the limitations in the adoption of both ECVs and ICCVs. These limitations are in terms of limited driving range and higher purchase cost for ECVs, and carbon emission limitations imposed by government regulations and Low Emission Zone (LEZ) penalties for ICCVs. This is a unique and complex model, and no study in the literature has addressed this problem sufficiently. A mathematical formulation and a heuristic algorithm are proposed for the defined time-dependent GVRP, and numerical experiments are designed and tested to demonstrate their capabilities. The results show that the proposed heuristic is efficient in finding sound solutions and can be used to identify changes in fleet design and routing of last-mile delivery operations as a result of green logistics policies.

## 7.1 Introduction

Traditionally, Vehicle Routing Problems (VRPs) aim to determine the low-cost routes for a fleet of vehicles to serve a set of customers such that the money and time spent to travel to those customer locations are minimized. However, the operation cost of routing vehicles is not limited to time and money. There are environmental costs associated with these operations that had almost been neglected before the twenty-first century until the substantial growth in Greenhouse

G. Masghati-Amoli (✉) · A. Haghani
University of Maryland, College Park, MD, USA
e-mail: golnush@umd.edu; haghani@umd.edu

© Springer Nature Switzerland AG 2020
H. Derbel et al. (eds.), *Green Transportation and New Advances in Vehicle Routing Problems*, https://doi.org/10.1007/978-3-030-45312-1_7

Gas (GHG) emissions and the world's climate change triggered the government awareness of the urgency to conserve the environment.

The most widely cited prediction of how the world's climate might change in the twenty-first century was made by the Working Groups of the Intergovernmental Panel on Climate Change (IPCC) in their 2nd Assessment in 1995. This report showed that during the twentieth century the average temperature of the world's climate had increased by 0.6 degrees centigrade, and it would increase by between 1.4 and 5.8 degrees centigrade by 2100 if there would be no changes in current human activities. On the other hand, the IPCC Fourth Assessment Report provided considerable evidence for the growing human influence on the climate system. It was shown that more than half of the observed increase in global average surface temperature from 1951 to 2007 was likely caused by the anthropogenic increase in GHG concentrations. The anthropogenic influences have likely contributed to retreat of glaciers since the 1960s and the increased surface melting of Greenland ice sheet since 1993. Transportation is one of the significant sources of anthropogenic GHG emissions. Based on the GHG emission reporting guidelines, transportation sector directly accounted for about 27% of the total U.S. GHG emissions in 2013, increased by 16% since 1990. Nearly, 97% of transportation GHG emissions came through direct combustion of fossil fuels with freight trucks playing as the third largest source of transportation GHG emissions. The increase in transportation-related GHGs is largely due to the increased demand for travel and an increase in the number of vehicle miles traveled by passenger cars and light-duty trucks due to population growth, economic growth, urban sprawl, and low fuel prices.

Following the substantial growth in GHG emissions, the government's awareness of the urgency to tackle these problems and conserve the environment increased, and green logistics received increased attention from governments and business organizations. The motivation behind green logistics was the unsustainable current production and distribution logistics strategies in the long term. As part of the green logistics program, carriers started to use Alternative Fuel Vehicles (AFVs), for last-mile deliveries to reduce their GHG emissions. In North America, large companies such as FedEx, General Electric, Coca-Cola, UPS, Frito-Lay, Staples, Enterprise, and Hertz (Electrification Coalition, 2013) started introducing battery–electric delivery vehicles to their last-mile delivery fleet. While this focus on truck conversion was desirable due to the large contribution of medium and large size trucks to GHG emission [31], many companies were still reluctant to adopt ECVs in their delivery fleet due to their high purchase cost and limited autonomy. Davis and Figliozzi [9] compared the whole life cost of battery–electric delivery trucks with that of a conventional ICCV serving less-than-truckload delivery routes, and the result of their analysis showed that the ECVs' total cost of operation was higher 86% of the times. It was concluded that a combination of factors such as high utilization rates, low speeds, and congestion, financial incentives, or technological breakthrough to reduce purchase cost would make ECVs a viable alternative to ICCVs. Taefi et al. [28] also argued that at this stage, incentives are likely to be needed to increase the commercial use of the BEVs. Government incentives could be in different forms from financial incentives such as subsidizing the purchase cost of ECVs and tax exemption incentives to providing access to inner-city areas with noise or pollution limits. Imposing emission caps on industry operations could also be another lever to promote the commercial use of BEVs.

The time-dependent green vehicle routing problem studied in this research is an extension of the classic vehicle routing problem aimed at controlling the environmental externalities associated with routing a mixed fleet of ECVs and ICCVs while addressing the limitations in the employment of both types of vehicles. The first study in Green Vehicle Routing Problem with a fleet of Alternative Fuel Vehicles was done by Erdogan and Miller-Hooks in 2012 [14]. Since then, different studies were conducted to improve the existing models by incorporating more realistic assumptions such as vehicle load capacity limitations, customer service time windows, mixed fleet of electric and combustion engine trucks, and different charging policies. While the models developed in previous studies were mainly focused on the limitations of electric trucks, in this research a more comprehensive GVRP was studied by accounting for the limitations and advantages associated with the use of both vehicle types. These limitations were defined in terms of limited range and higher purchase cost for ECVs, and the emission limitations imposed by the government and the Low Emission Zone penalties for ICCVs.

Moreover, due to the important role of travel time on vehicle energy requirements, a time-dependent GVRP was formulated in this research. Accounting for travel time variations enabled the model to account for varying congestion level on each arc in finding optimal routes. This enhanced the reliability of the model, especially in routing ECVs as electric vehicle routes are very sensitive to energy consumption estimations because of their limited driving range. Also, contrary to previous studies, a heterogeneous fleet of ECVs and ICCVs with different battery and load capacities were considered in the problem. The route plans were not determined for a pre-specified number of vehicles. It was assumed that the number of vehicles is not predefined, and the optimal number of vehicles of each type is to be found.

## 7.2   Literature Review

The research on Green Vehicle Routing Problem (GVRP) deals with the optimization of energy consumption and, as a result, pollution of logistics transportation activities. It deals with the decisions related to routing and scheduling of vehicles, and the choice of vehicle type for given deliveries, particularly concerning the potential added cost of $CO_2$ emissions. Studies on Green Vehicle Routing Problem can be divided into two categories. The first category is the VRP in which fuel consumption and emissions of combustion engine vehicles are minimized as part of their routing costs. This type of GVRP is called Energy Minimizing Vehicle Routing Problem (EMVRP) or Pollution Routing Problem (PRP). The Energy Minimizing Vehicle Routing Problem (EMVRP) integrates the cost of vehicle energy consumption into the routing cost of the classic vehicle routing problem.

Therefore, not only the energy consumption is minimized, but also the decrease in petroleum-based fuel consumption reduces greenhouse gas emissions. One of the important components of the EMVRP is the energy consumption model used to estimate the vehicle energy requirements for routing. A more accurate energy consumption model results in a more realistic estimation of the vehicle energy requirements. The EMVRPs studied in the literature vary from each other in terms of the factors they consider in their energy consumption models. According to the report by the US Department of Energy (2008), travel speed, vehicle load, and transportation distance are among the significant factors affecting vehicle fuel consumption. Moreover, the results of the studies by Ardekani et al. [2], Bigazzi and Bertini [6], Demir et al. [10], and Alwakiel [1] show that vehicle characteristics, environment, and traffic conditions, and driver behavior are significant contributors to vehicle energy consumption. Most of the used models in the EMVRP studies concentrate on the vehicle, traffic, and environmental aspects and do not capture driver-related factors that are relatively difficult to measure. The EMVRP was first introduced by Kara et al. in 2007[19]. They defined an EMVRP as a capacitated VRP with the objective of minimizing the routing cost in the form of a weighted load function, defined as the product of total vehicle load and arc length. Another study on the effect of vehicle load on energy consumption was done by Xiao et al. in 2012[32]. In this study, the authors modeled fuel consumption as a linear function of vehicle load and distance. It was concluded that the shortest distance might not be the optimal solution to lower fuel consumption because distance and vehicle load contribute to the total fuel consumption jointly. While these studies mainly focus on vehicle load in modeling vehicle energy consumption, there are several other studies that consider speed as another contributing factor to energy consumption besides distance and load. Eglese and Black [13] showed that speed is a more important factor than distance when estimating fuel consumption and emissions. Kuo [20] solved the time-dependent VRP that aimed to minimize fuel consumption as a function of vehicle speed varying over different times of day. The study findings showed that the proposed method provided a 24.61% improvement in fuel consumption over the method based on minimizing transportation time and a 22.69% improvement over the method based on minimizing transportation distances. In another study by Maden [23], it was concluded that the standard time-dependent VRP with the objective of minimizing total travel time results in a saving in fuel consumption as the model avoids congested links to minimize the travel time. Their study results showed an average of 7% reduction in the vehicles fuel consumption. Jovicic et al.'s [18] investigations showed that in the City of Kragujevac in Serbia, a reduction of up to 20% can be achieved in energy costs and the associated emissions if the effect of vehicle speed on the vehicle energy consumption is accounted for in the routing of the municipal waste collection. Contrary to these studies in which the vehicle speed is given, Bektas and Laporte [5] solved a capacitated VRP with time windows with speed on each arc as a primary decision variable to find the optimal speed for vehicle movements along each arc such that the total energy consumption is minimized. The energy consumption in this study was formulated as a function of vehicle load and travel speed. In another research done by Demir et al.[11], the trade-offs between fuel consumption and driving time were investigated. They showed that trucking companies need not compromise greatly in terms of driving time to achieve a significant reduction in fuel consumption and $CO_2$ emissions. They also argued that the converse of this insight holds too and considerable reductions in driving time are achievable if one is willing to increase fuel consumption only slightly. Tavares et al. [29] looked at an EMVRP optimizing the routing cost of waste transportation by taking into account the energy consumption as a function of road angle besides vehicle load. Their findings indicated that optimizing fuel consumption can yield savings of up to 52% in fuel when compared with minimizing distance. In another paper by Tavares et al. [30], the routing of municipal solid waste collection fleet was optimized by minimizing the fuel consumption using 3D GIS modeling with road gradient playing the main role in energy requirement considerations. The results of this study suggested that the proposed methodology reduced traveled distance and fuel consumption by 29% and 16%, respectively. The most comprehensive energy model was used in Demir et al. [10] study where they compared several energy consumption models and revealed other relevant contributing factors such as driver acceleration behavior, engine type, and size, vehicle design, besides road gradient, speed, and vehicle load.

The second category of GVRPs deals with energy consumption and also limitations of routing a fleet of alternative fuel vehicles. This type of GVRP is called Alternative Fuel Vehicle Routing Problem (AFVRP), in general, or Electric Vehicle Routing Problem (EVRP) in particular when the fleet of AFVs is of the battery–electric type. AFVRP is closely related to classical distance-constrained VRP [21]; however, in AFVRP, there is a possibility of extending the vehicle's distance limitation by visiting charging stations. Therefore, existing solutions to distance-constrained VRP could not be applied to AFVRP. Another type of problem that is closely related to AFVRP is the multi-depot vehicle routing problem with inter-depot facilities described by Bard et al. [4]. This problem considers intermediate depots at which vehicles can be reloaded to serve customer demands. In the literature, relatively few studies have been published on alternative fuel vehicle optimization problems. The paper by Erdogan and Miller-Hooks [14] was one of the pioneers in AFVRP studies. In this study, an AFVRP model was developed to optimally route a fleet of un-capacitated AFVs with a limited driving range and the possibility of refueling at dedicated stations having unlimited capacity. Two heuristics were proposed to find the tours with minimum total distance while eliminating the risk of running out of fuel. It was assumed that the vehicles are fully charged upon each visit to charging stations with a constant charging time.

In 2013, Barco et al. [3] expanded Erdogan and Miller-Hooks study by considering more realistic assumptions such as vehicle load capacity. They tried to find optimal routes for a set of homogeneous capacitated electric airport shuttles by minimizing total energy consumption, recharging, and battery degradation costs. Scheduling of charges was coordinated with routing to guarantee a reliable operation serving the demand of customers within their time windows while accounting for the variation of the energy cost during the peak and non-peak hours of the day.

In a study done by Schneider et al. in 2014[26], a different variant of capacitated AFVRP with customer time windows was solved in which the optimal number of required vehicles had to be found as well as the minimum total distance tours using a hierarchical objective function. All available electric vehicles were assumed to be homogeneous, and the charging time was assumed to vary depending on the state of the battery upon arrival to charging stations.

Felipe et al. [15] extended Erdogan and Miller-Hooks study in an alternative way by including realistic considerations such as the possibility of performing a partial recharge at a station and the availability of different charging technologies, implying different recharging times and costs (slow, fast, and wireless). The variability in charging policies used in previous studies motivated Desaulniers et al. in 2014[12] to investigate the effect of these different charging policies on the total cost of a fleet of electric delivery trucks routing to distribute goods to customers with soft service time windows. The result showed that allowing multiple and partial charges along the route for each electric truck helps to reduce the routing cost and the number of employed vehicles in comparison to the variants with single and full charges. The results from Bruglieri et al. study in 2015[7] showed the same findings. They proposed a Mixed Integer Linear Programming formulation of the EVRP with time windows to minimize the total travel, waiting and recharging time plus the number of the employed EVs assuming partial charging multiple stops is allowed. Their findings showed that partial charging policy outperforms the full charging one in terms of recharging, waiting, and travel time. While all these studies were focusing on a homogeneous fleet of AFVs, Hiermann et al. in 2014[17] expanded the AFVRP with time windows by considering a mixed fleet of fixed size electric vehicles with different battery capacities, load capacities, and purchase costs. The charging policy considered in this study was the single and full charge in each route.

Considering the fact that most companies do not operate pure EV fleets and are gradually introducing ECVs into their existing internal combustion engine vehicle fleet, Sassi et al. [25] tried to model and solve an EVRP with a mixed fleet of heterogeneous EVs and homogeneous combustion engine commercial vehicles. The objective of this study was to find the minimum cost routes while accounting for limited range of EVs. It was assumed that the charging cost is dependent on the time of a day, and specific working hours were assumed for charging stations. Moreover, it was assumed that EVs could only charge at charging stations with compatible charging techniques with partial charging allowed. A greener version of Mixed Fleet EVRP was studied by Goekea and Schneider [16]. The developed model in their study was aimed to minimize vehicle energy consumption cost. The vehicle energy requirements were modeled as a function of travel speed, gradient, and cargo load, and contrary to Sashi's study, only full charge of the EVs was allowed upon each visit to charging stations. The result of their study showed that consideration of the vehicle load in the EV battery consumption estimation model strongly improves the quality of the generated routing solutions. Moreover, it was found that a large number of solutions that are generated without load estimates are actually infeasible due to battery capacity.

In this research, a more comprehensive GVRP was defined to address the gaps in the literature. First of all, due to the sensitivity of ECV's driving range to the energy consumption rate, a time-dependent GVRP was introduced. By accounting for time dependency of travel time along the arcs in the network, recurring congestion was taken into account providing a more realistic estimation of energy requirement by vehicles while routing. Moreover, while all the studies in the literature only focus on the limitations of ECV adoption, in this research, a new problem was defined where there are limitations in the adoption of both types of vehicles. These limitations were limited driving range and high acquisition cost for ECVs and LEZ penalty and emission cap for ICCVs. Contrary to previous studies, a heterogeneous fleet of ECVs and ICCVs with different battery and load capacities were considered in the problem. The route plans were not determined for a pre-specified number of vehicles, and it was assumed that the optimal fleet design is to be found.

## 7.3  Mathematical Model Development

In the literature, different approaches have been used to take into account the time dependency of travel time while finding solutions to variants of VRP. Some studies used step functions to represent the variation of travel time along different time periods of a day. While using a step function helps to account for variations in travel time, it might fail to take into account the First In First Out (FIFO) concept. Based on the FIFO concept, if two vehicles leave from the same location for the same destination traveling on the same path, the one that leaves first will always arrive first, no matter how speed changes along the arcs during the travel. However, as it can be seen in Fig. 7.1, if it is assumed that travel time is a step function of departure time, the vehicles departing at 11:01 are expected to arrive after the vehicles departing at 10:55, which violates the FIFO concept.

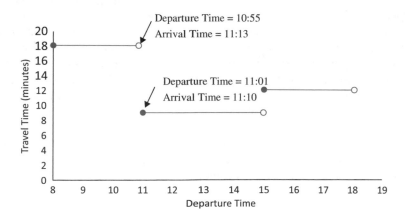

**Fig. 7.1**  Travel time step function

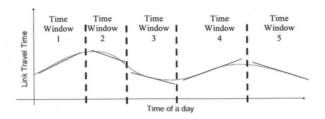

**Fig. 7.2** Travel time as a continuous function of time of a day

To preserve the FIFO concept, a continuous function was used in this study to represent variation in travel time during different times of a day, as shown in Fig. 7.2. Therefore, any type of variations in travel time is accepted and taken into account by the model developed in this study. As can be seen in Fig. 7.2, travel time can be a nonlinear function of time. Therefore, piecewise linear functions were used to estimate travel time as a linear function of time during different time windows. Each time window $t$ was represented by its range as $[LT^t, UT^t]$ and its travel time function in the form of:

$$tt^t = intercept^t + slope^t \times (DepartureTime - LT^{(t-1)}) \quad \forall t \in \{1, 2, \ldots, n\} \tag{7.1}$$

where $intercept^t$ and $slope^t$ are the intercept and slope of the fitted line at time window $t$, $LT^t$ is the lower bound of time window $t$.[1]

A nonnegative demand, $q_i$, a nonnegative service time, $s_i$, and a soft service time window $[e_i, l_i]$ were associated with each demand node in the network. If the demands were not served on time, a delay or waiting penalty was imposed. The charging stations were assumed to be of battery swapping type with no limitation on operating hours. Once an ECV was reached to a charging station, its battery was swapped with a fully charged battery. Therefore, the charging time was assumed to be constant for all types of ECVs at all charging station locations.

It was assumed that (a) there is a financial incentive by government subsidizing the purchase of ECVs, (b) there is a number of Low Emission Zones (LEZ) in the network operating 24 h a day and 365 days of a year such that combustion engine commercial vehicles are required to pay a daily charge to drive within these zones, and (c) there is an emission cap imposed by the government on the amount of pollution that a company can produce in a year.

A mixed fleet of heterogeneous electric commercial vehicles, ECVs, with different battery and loading capacities, and heterogeneous Internal Combustion Engine Vehicles, ICCVs, with different loading capacities was considered to be available in this study. The energy requirement of vehicles was estimated based

---

[1]It should be noted that in this case, the FIFO concept is guaranteed only if the absolute value of the slope of this linear function is always less than 1.

on mechanical power, $P_M$, required by the vehicle to overcome rolling resistance, aerodynamic resistance, and gravitational force.

$$P_M = \left( m.a + \frac{1}{2}.c_d.\rho.A.v^2 + m.g.\sin(z) + c_r.m.g.\cos(z) \right).v \qquad (7.2)$$

where $m$ is the vehicle mass defined as the sum of vehicle weight, $W$, and vehicle load, $U$; $a$ is the acceleration rate; $C_d$ is the aerodynamic drag coefficient; $\rho$ is the air density; $A$ is the frontal area of the vehicle; $C_r$ is the rolling friction coefficient; $v$ is the vehicle speed; and $z$ is the road gradient.

Having estimated the required mechanical power, energy requirement for ECVs and ICCVs was estimated using equations 3 and 4, respectively.

$$P_B = \varphi.P_M \qquad (7.3)$$

where $P_B$ is the required battery energy per second, and $\varphi$ is the regression coefficient that describes the battery efficiency.

$$FR = \frac{\xi}{K.\psi} \left( KND + \frac{P_M}{\eta.\eta_{tf}} \right) \qquad (7.4)$$

where $FR$ is the fuel consumption rate per second, $\xi$ is the fuel-to-air mass ratio, $k$ is the heating value of typical diesel fuel, $K$ is the engine friction factor, $N$ is the engine speed, $D$ is the engine displacement, $\psi$ is a factor converting the fuel rate from grams per second to liters per second, $\eta$ is an efficiency parameter for diesel engines, and $\eta_{tf}$ is the drive train efficiency.

Since the instantaneous engine-out Greenhouse Gas emission rate is directly related to the fuel consumption rate, $FR$, [5], the vehicle emission was estimated in gram per second (g/s) through equation 5.

$$E = \delta_1 FR + \delta_2 \qquad (7.5)$$

where $\delta_1$ and $\delta_2$ are GHG-specific emission index parameters.

### 7.3.1  Mathematical Formulation

The time-dependent GVRP of interest was formulated as a mixed integer linear programming problem based on the above assumptions on a complete directed graph $G = (V'_{(0,N+1)}, A)$. Each arc in the network was described by its travel time as a function of departure time, average travel speed ($V^t_{ij}$), and the average acceleration rate ($a^t_{ij}$) for different time periods of a day. The notations, coefficients, and variables used in the mathematical model are introduced in Table 7.1.

**Table 7.1** Notation and variables

| Notation\Variables | Definition |
| --- | --- |
| $V$ | Set of demand nodes |
| $F'$ | Set of charging station visit nodes, dummy vertices of the set of charging stations $F$ |
| $V'$ | $V \cup F'$ |
| $V_0, V_{N+1}$ | Instances of depot |
| $V_{0,N+1}$ | $V \cup V_0 \cup V_{N+1}$ |
| $V'_{0,N+1}$ | $V' \cup V_0 \cup V_{N+1}$ |
| $LEZ_l$ | Set of demand nodes in the Low Emission Zone $l \in 1, 2, \ldots, L$ |
| $C_{CV}$ | Set of ICCV types |
| $C_{EV}$ | Set of ECV types |
| $q_i$ | Nonnegative demand of node $i \in V$ |
| $e_i$ | Earliest service time of node $i \in V$ |
| $l_i$ | Latest service time of node $i \in V$ |
| $S_i$ | Service time of node $i \in V'$ |
| $FC_c^{EV}$ | Fixed cost of EV type $c \in C_{EV}$ (\$/day) |
| $FC_c^{CV}$ | Fixed cost of ICCV type $c \in C_{CV}$ (\$/day) |
| $Q_c^{EV}$ | Loading capacity of EV type $c \in C_{EV}$ |
| $Q_c^{CV}$ | Loading capacity of ICCV type $c \in C_{CV}$ |
| $W_c^{EV}$ | Weight of empty EV of type $c \in C_{EV}$ |
| $W_c^{CV}$ | Weight of empty ICCV of type $c \in C_{CV}$ |
| $BC_c^{EV}$ | Battery capacity of EV type $c \in C_{EV}$ |
| $C^E$ | Cost of electricity (\$/kw) |
| $C^F$ | Cost of fuel (\$/liter) |
| $C^L$ | Cost of labor |
| $v_{ij}^t$ | Average travel speed on arc $(i, j)|(i, j) \in V'_{0,N+1}, i \neq j$, at time interval $t \in T$ |
| $a_{ij}^t$ | Average acceleration rate on arc $(i, j)|(i, j) \in V'_{0,N+1}, i \neq j$, at time interval $t \in T$ |
| $WP$ | Waiting time penalty |
| $DP$ | Delay time penalty |
| $P$ | LEZ daily penalty |
| $e^{CAP}$ | Emission Cap |
| $P_{Carb}$ | Price of carbon (\$/gram) |
| $PB$ | Battery cycle cost |
| $ghg$ | Greenhouse gas emissions per liter of fuel (gram/liter) |
| $LT_t$ | The lower boundary of time period $t \in T$ |
| $UT_t$ | The upper boundary of time period $t \in T$ |

(continued)

**Table 7.1** (continued)

| Notation\Variables | Definition |
|---|---|
| $LU_m$ | The lower boundary of load interval $m \in M$ |
| $UU_m$ | The upper boundary of load interval $m \in M$ |
| $\overline{U}_m$ | Average of load in load interval $m \in M$ |
| $T_0, T_{End}$ | Start and End time of delivery operations |
| $x_{ijcktm}^{EV}$ | Binary variable with value of 1 if ECV $k$ of type $c$ travels from node $i$ to node $j$ in time interval $t$ carrying load in interval $m$ |
| $x_{ijcktm}^{CV}$ | Binary variable with value of 1 if ICCV $k$ of type $c$ travels from node $i$ to node $j$ in time interval $t$ carrying load in interval $m$ |
| $y_{ckl}^{CV}$ | Binary variable with value of 1 if engine vehicle $k$ of type $c$ enters Low Emission Zone $l$ |
| $at_{ick}^{EV}$ | Arrival time of electric vehicle $k$ of type $c$ at node $i$ |
| $at_{ick}^{CV}$ | Arrival time of combustion engine vehicle $k$ of type $c$ at node $i$ |
| $dt_{ick}^{EV}$ | The departure time of electric vehicle $k$ of type $c$ from node $i$ |
| $dt_{ick}^{CV}$ | The departure time of combustion engine vehicle $k$ of type $c$ from node $i$ |
| $DT_{ijcktm}^{EV}$ | The departure time of ECV $k$ of type $c$ from node $i$ to node $j$ in time interval $t$ carrying load in interval $m$ |
| $DT_{ijcktm}^{CV}$ | The departure time of ICCV $k$ of type $c$ from node $i$ to node $j$ in time interval $t$ carrying load in interval $m$ |
| $U_{ijck}^{EV}$ | The load carried by electric vehicle $k$ of type $c$ from node $i$ to node $j$ |
| $U_{ijck}^{CV}$ | The load carried by combustion engine vehicle $k$ of type $c$ from node $i$ to node $j$ |
| $R_{ick}$ | The remaining battery of electric vehicle $k$ of type $c$ upon arrival at node $i$ |
| $w_{ick}^{EV}$ | Waiting time of electric vehicle $k$ of type $c$ at node $i$ |
| $w_{ick}^{CV}$ | Waiting time of combustion engine vehicle $k$ of type $c$ at node $i$ |
| $d_{ick}^{EV}$ | Delayed time of electric vehicle $k$ of type $c$ at node $i$ |
| $d_{ick}^{CV}$ | Delayed time of combustion engine vehicle $k$ of type $c$ at node $i$ |

### 7.3.1.1 Objective Function

The objective is to minimize the total operation cost that includes vehicle purchase cost, fuel and electric energy consumption cost, labor cost, and the total penalty cost of LEZ, service time, and emission. The formulation for the objective function is presented in Eq. 7.6.

The first and second rows of the objective function estimate vehicle purchase costs, where $F_c^{EV}$ and $F_c^{CV}$ are the fixed cost of ECV and ICCV per day of operation. The fuel energy consumed by ICCVs is minimized in rows 3–5, followed by the electric energy consumption of ECVs in rows 6–8. The ninth row of objective function minimizes the LEZ penalty cost imposed if any ICCV enters any LEZ in the network. The user inconvenience cost is minimized in the tenth and eleventh rows of the objective function in the form of waiting or delayed service cost penalties. In row twelve, emission cost is minimized by either decreasing the cost of emission produced more than the limit or by increasing the amount of extra emission credit to be sold to other companies. The labor cost is minimized in the fifteenth row. The final row minimizes the battery degradation cost, which is calculated as the multiplication of the battery cycle cost by the number of times the battery of ECV is replaced at a charging station. The battery cycle cost is estimated by dividing the cost of battery by the number of times it can be fully charged in its life cycle.

$Minimize:$

$$\sum_{(c \in C^{EV})} \sum_{\left(k \in S_c^{EV}\right)} \sum_{(j \in V')} \sum_{(t \in T)} \sum_{(m \in M)} F_c^{EV} \cdot x_{0jcktm}^{EV} +$$

$$\sum_{(c \in C^{CV})} \sum_{\left(k \in S_c^{CV}\right)} \sum_{(j \in V)} \sum_{(t \in T)} \sum_{(m \in M)} F_c^{CV} \cdot x_{0jcktm}^{CV} +$$

$$C^F \times \sum_{(c \in C^{CV})} \sum_{\left(k \in S_c^{CV}\right)} \sum_{(j \in V)} \sum_{(t \in T)} \left( (\alpha_{ij} \times w_c^{CV} + \beta_{ij}) \times \left( intercept_{ij}^t \times \sum_m x_{ijcktm}^{CV} + \right.\right.$$

$$slope_{ij}^t \times \sum_m \left( DT_{ijcktm}^{CV} - LT_t \times x_{ijcktm}^{CV} \right) \bigg) +$$

$$\alpha_{ij} \times \left( intercept_{ij}^t \times \sum_m M_{avg} \times x_{ijcktm}^{CV} + slope_{ij}^t \times \sum_m \overline{U}_m \right.$$

$$\left.\left.\times \left( DT_{ijcktm}^{CV} - LT_t \times x_{ijcktm}^{CV} \right) \right) \right) +$$

$$C^E \times \sum_{(c \in C^{EV})} \sum_{\left(k \in S_c^{EV}\right)} \sum_{(j \in V)} \sum_{(t \in T)} \left( \left(\alpha_{ij} \times w_c^{EV} + \beta_{ij}\right) \times \left( intercept_{ij}^t \times \sum_m x_{ijcktm}^{EV} + \right.\right.$$

$$slope_{ij}^t \times \sum_m \left( DT_{ijcktm}^{EV} - LT_t \times x_{ijcktm}^{EV} \right) \bigg) +$$

$$\alpha_{ij} \times \left( intercept_{ij}^t \times \sum_m M_{avg} \times x_{ijcktm}^{EV} + slope_{ij}^t \times \sum_m \overline{U}_m \times \right.$$

$$\left( DT_{ijcktm}^{EV} - LT_t \times x_{ijcktm}^{EV} \right) \bigg) \bigg) + P \times \sum_{(c \in C^{CV})} \sum_{\left(k \in S_c^{CV}\right)} \sum_{(l \in L)} y_{kl}^{CV} +$$

$$WP \times \sum_{i \in V} \sum_{c \in C^{EV}} \sum_{k \in S_c^{EV}} w_{ick}^{EV} + \sum_{i \in V} \sum_{c \in C^{CV}} \sum_{k \in S_c^{CV}} w_{ick}^{CV} +$$

$$DP \times \sum_{i \in V} \sum_{c \in C^{EV}} \sum_{k \in S_c^{EV}} d_{ick}^{EV} + \sum_{i \in V} \sum_{c \in C^{CV}} \sum_{k \in S_c^{CV}} d_{ick}^{CV} -$$

$$
P_{carb} \times \left( e^{CAP} - ghg \times \left[ \sum_{c \in C^{CV}} \sum_{k \in S_c^{CV}} \sum_{j \in V} \sum_{t \in T} \left( \left( \alpha_{ij} \times w_c^{CV} + \beta_{ij} \right) \times \right. \right. \right.
$$

$$
\left( intercept_{ij}^t \times \sum_m x_{ijcktm}^{CV} + slope_{ij}^t \times \sum_m \left( DT_{ijcktm}^{CV} - LT_t \times x_{ijcktm}^{CV} \right) \right) +
$$

$$
\alpha_{ij} \times \left( intercept_{ij}^t \times \sum_m M_{avg} \times x_{ijcktm}^{CV} + \right.
$$

$$
\left. \left. \left. slope_{ij}^t \times \sum_m \overline{U}_m \times \left( DT_{ijcktm}^{CV} - LT_t \times x_{ijcktm}^{CV} \right) \right) \right) \right] \right) +
$$

$$
C^L \times \left( \sum_{c \in C^{EV}} \sum_{k \in S_c^{EV}} at_{N+1ck}^{EV} + \sum_{c \in C^{CV}} \sum_{k \in S_c^{CV}} at_{N+1ck}^{CV} \right) +
$$

$$
PB \times \sum_{i \in V_0} \sum_{j \in F'} \sum_{c \in C^{EV}} \sum_{k \in S_c^{EV}} \sum_{t \in T} \sum_{m \in M} x_{ijcktm}^{EV} \qquad (7.6)
$$

### 7.3.1.2 Constraints

In the developed mathematical formulation, the general constraints of vehicle routing problem were extended to represent the variant of VRP defined in this study. Constraint 7.7 ensures that each demand node has exactly one successor. Constraints 7.8 and 7.9 guarantee that for each node in the network, the number of incoming arcs is equal to the number of outgoing arcs for each vehicle type. Constraints 7.10 and 7.11 force each vehicle to be assigned to a maximum of one route. Constraint 7.12 ensures that each node in the set $F'$ is visited at most once by each vehicle. This makes it possible for each charging station to be visited once, multiple times, or not at all by the ECVs on the road. As mentioned in Table 7.1, $F'$ is a set of dummy nodes representing visits to each vertex in the set of charging stations, $F$.

$$
\sum_{c \in C_{EV}} \sum_{k \in K_{EV}} \sum_{j \in V'_{N+1}} \sum_{t \in T} \sum_{m \in M} x_{ijcktm}^{EV} + \sum_{c \in C_{CV}} \sum_{k \in K_{CV}} \sum_{j \in V_{N+1}} \sum_{t \in T} \sum_{m \in M} x_{ijcktm}^{CV} = 1
$$

$$
\forall i \in V, i \neq j
$$
$$
(7.7)
$$

$$
\sum_{j \in V'_{N+1}} \sum_{t \in T} \sum_{m \in M} x_{ijcktm}^{EV} - \sum_{j \in V'_0} \sum_{t \in T} \sum_{m \in M} x_{jicktm}^{EV} = 0
$$

$$
\forall i \in V', c \in C_{EV}, k \in K_{EV}, i \neq j \qquad (7.8)
$$

$$\sum_{j \in V_{N+1}} \sum_{t \in T} \sum_{m \in M} x_{ijcktm}^{CV} - \sum_{j \in V_0} \sum_{t \in T} \sum_{m \in M} x_{jicktm}^{CV} = 0$$

$$\forall i \in V, c \in C_{CV}, k \in K_{CV}, i \neq j \tag{7.9}$$

$$\sum_{j \in V'} \sum_{t \in T} \sum_{m \in M} x_{0jcktm}^{EV} \leq 1 \quad \forall c \in C_{EV}, k \in K_{EV} \tag{7.10}$$

$$\sum_{j \in V} \sum_{t \in T} \sum_{m \in M} x_{0jcktm}^{CV} \leq 1 \quad \forall c \in C_{CV}, k \in K_{CV} \tag{7.11}$$

$$\sum_{j \in V_{n+1}} \sum_{t \in T} \sum_{m \in M} x_{ijcktm}^{EV} \leq 1 \quad \forall c \in C_{EV}, k \in K_{EV}, i \in F', i \neq j \tag{7.12}$$

$$at_{jck}^{EV} \geq at_{ick}^{EV} + S_i \sum_{t \in T} \sum_{m \in M} x_{ijcktm}^{EV} + \sum_{t \in T} \sum_{m \in M} intercept_{ij}^t \times x_{ijcktm}^{EV}$$

$$+ \sum_{t \in T} \sum_{m \in M} Slope_{ij}^t \times \left( DT_{ijcktm}^{EV} - LT_{t-1} \times x_{ijcktm}^{EV} \right)$$

$$- T. \left( 1 - \sum_t \sum_m x_{ijcktm}^{EV} \right) \quad \forall c \in C_{EV}, k \in K_{EV}, i \in V_0', j \in V_{n+1}', i \neq j$$

$$\tag{7.13}$$

$$at_{jck}^{CV} \geq at_{ick}^{CV} + S_i \sum_{t \in T} \sum_{m \in M} x_{ijcktm}^{CV} + \sum_{t \in T} \sum_{m \in M} intercept_{ij}^t \times x_{ijcktm}^{CV}$$

$$+ \sum_{t \in T} \sum_{m \in M} Slope_{ij}^t \times \left( DT_{ijcktm}^{CV} - LT_{t-1} \times x_{ijcktm}^{CV} \right)$$

$$- T. \left( 1 - \sum_t \sum_m x_{ijcktm}^{CV} \right) \quad \forall c \in C_{CV}, k \in K_{CV}, i \in V_0, j \in V_{n+1}, i \neq j$$

$$\tag{7.14}$$

Travel times on arcs are linked through constraints 7.13 and 7.14, which ensure the connectivity of travel times on the traveled arcs by each vehicle. Basically, arrival time at node $j$ for an ECV can be formulated as Eq. 7.15, where $tt_{ij}^t$ is the travel time on arc $(i, j)$, which is calculated based on Eq. 7.16. Since $dt_{ick}^{EV}$ is a decision variable, multiplying $tt_{ij}^t$ by $x_{ijcktm}^{EV}$ makes the problem nonlinear. In order to avoid nonlinearity, $(tt_{ij}^t + S_i) \cdot \sum_t \sum_m x_{ijcktm}^{EV}$ in Eq. 7.15 is substituted in Eq. 7.17, which results in the formulation represented in constraints 7.13 and 7.14.

$$at_{jck}^{EV} \geq at_{ick}^{EV} + \left(tt_{ij}^t + S_i\right) \cdot \sum_t \sum_m x_{ijcktm}^{EV} - T \cdot \left(1 - \sum_t \sum_m x_{ijcktm}^{EV}\right)$$

$$\forall c \in C_{EV}, k \in K_{EV}, i \in V_0', j \in V_{n+1}', i \neq j \tag{7.15}$$

$$tt_{ij}^t = intercept_{ij}^t + Slope_{ij}^t \times \left(dt_{ick}^{EV} - LT_t\right) \tag{7.16}$$

$$S_i \sum_{t \in T} \sum_{m \in M} x_{ijcktm}^{EV} + \sum_{t \in T} \sum_{m \in M} intercept_{ij}^t \times x_{ijcktm}^{EV}$$

$$+ \sum_{t \in T} \sum_{m \in M} Slope_{ij}^t \times \left(DT_{ijcktm}^{EV} - LT_{t-1} \times x_{ijcktm}^{EV}\right) \tag{7.17}$$

Departure time from each node in the network is calculated through constraints 7.18 and 7.19. The estimated departure time is then used to estimate the decision variables $DT_{ijcktm}^{CV}$ and $DT_{ijcktm}^{EV}$ as shown in constraints 7.20 through 7.23. Constraints 7.24–7.27 guarantee that the arrival time and departure time of vehicles at each node are within the daily delivery operations working hour $[T_0, T]$. The departure time of each vehicle is associated with the corresponding time window $t$ through constraints 7.28–7.31. The identified time window is then used to select the proper travel time function for travel time estimation.

$$dt_{ick}^{EV} \geq at_{ick}^{EV} + S_i \cdot \sum_{j \in V_{N+1}'} \sum_t \sum_m x_{ijcktm}^{EV} \qquad \forall c \in C_{EV}, k \in K_{EV}, i \in V'$$

$$\tag{7.18}$$

$$dt_{ick}^{CV} \geq at_{ick}^{CV} + S_i \cdot \sum_{j \in V_{N+1}} \sum_t \sum_m x_{ijcktm}^{CV} \qquad \forall c \in C_CV, k \in K_{CV}, i \in V$$

$$\tag{7.19}$$

$$DT_{ijcktm}^{EV} \leq M x_{ijcktm}^{EV} \qquad \forall c \in C_{EV}, k \in K_{EV}, i \in V', j \in V_{N+1}', t \in T, m \in M$$

$$\tag{7.20}$$

$$\sum_{j \in V_{N+1}} \sum_t \sum_m DT_{ijcktm}^{EV} = dt_{ick}^{EV} \qquad \forall c \in C_{EV}, k \in K_{EV}, i \in V' \tag{7.21}$$

$$DT_{ijcktm}^{CV} \leq M x_{ijcktm}^{CV} \qquad \forall c \in C_{CV}, k \in K_{CV}, i \in V, j \in V_{N+1}, t \in T, m \in M$$

$$\tag{7.22}$$

$$\sum_{j \in V_{N+1}} \sum_t \sum_m DT_{ijcktm}^{CV} = dt_{ick}^{CV} \qquad \forall c \in C_{CV}, k \in K_{CV}, i \in V \tag{7.23}$$

$$T_0 \leq at_{jck}^{EV} \leq T \qquad \forall c \in C_{EV}, k \in K_{EV}, j \in V_{n+1}' \tag{7.24}$$

$$T_0 \le at_{jck}^{CV} \le T \quad \forall c \in C_{CV}, k \in K_{CV}, j \in V_{n+1} \tag{7.25}$$

$$T_0 \le dt_{jck}^{EV} \le T \quad \forall c \in C_{EV}, k \in K_{EV}, j \in V'_{n+1} \tag{7.26}$$

$$T_0 \le dt_{jck}^{EV} \le T \quad \forall c \in C_{CV}, k \in K_{CV}, j \in V_{n+1} \tag{7.27}$$

$$dt_{ick}^{CV} - UT_t \le 100 \left(1 - \sum_m x_{ijcktm}^{CV}\right)$$

$$\forall c \in C_{CV}, k \in K_{CV}, i \in V_0, j \in V_{n+1}, t \in T \tag{7.28}$$

$$dt_{ick}^{CV} \ge LT_t \sum_m x_{ijcktm}^{CV} \quad \forall c \in C_{CV}, k \in K_{CV}, i \in V_0, j \in V_{n+1}, t \in T$$

$$\tag{7.29}$$

$$dt_{ick}^{EV} - UT_t \le 100 \left(1 - \sum_m x_{ijcktm}^{EV}\right)$$

$$\forall c \in C_{EV}, k \in K_{EV}, i \in V'_0, j \in V'_{n+1}, t \in T \tag{7.30}$$

$$dt_{ick}^{EV} \ge LT_t \sum_m x_{ijcktm}^{EV} \quad \forall c \in C_{EV}, k \in K_{EV}, i \in V'_0, j \in V'_{n+1}, t \in T$$

$$\tag{7.31}$$

If the arrival of a vehicle at a demand node is not within the desired delivery time window, early or delayed service time at that demand node is calculated using constraints 7.32–7.35. The total load a vehicle carries is limited by its capacity through constraints 7.36 and 7.37. Constraints 7.38 and 7.39 force the load carried by a vehicle over an arc $(i, j)$ to be zero if that vehicle does not traverse the arc.

$$w_{jck}^{EV} \ge e_j - at_{jck}^{EV} \quad \forall c \in C_{EV}, k \in K_{EV}, j \in V \tag{7.32}$$

$$d_{jck}^{EV} \ge at_{jck}^{EV} - l_j \quad \forall c \in C_{EV}, k \in K_{EV}, j \in V \tag{7.33}$$

$$w_{jck}^{CV} \ge e_j - at_{jck}^{CV} \quad \forall c \in C_{CV}, k \in K_{CV}, j \in V \tag{7.34}$$

$$d_{jck}^{CV} \ge at_{jck}^{CV} - l_j \quad \forall c \in C_{CV}, k \in K_{CV}, j \in V \tag{7.35}$$

$$\sum_{i \in V} q_i \sum_{j \in V_{N+1}} \sum_t \sum_m x_{ijcktm}^{EV} \le Q_c^{EV} \quad \forall c \in C_{EV}, k \in K_{EV} \tag{7.36}$$

$$\sum_{i \in V} q_i \sum_{j \in V_{N+1}} \sum_t \sum_m x_{ijcktm}^{CV} \le Q_c^{EV} \quad \forall c \in C_{CV}, k \in K_{CV} \tag{7.37}$$

$$q_j \cdot \sum_t \sum_m x_{ijcktm}^{EV} \le U_{ijck}^{EV} \le (Q_c^{EV} - q_i) \cdot \sum_t \sum_m x_{ijcktm}^{EV}$$

$$\forall c \in C_{EV}, k \in K_{EV}, i \in V_0, j \in V_{n+1} \tag{7.38}$$

$$q_j \cdot \sum_t \sum_m x_{ijcktm}^{CV} \le U_{ijck}^{CV} \le (Q_c^{CV} - q_i) \cdot \sum_t \sum_m x_{ijcktm}^{CV}$$

$$\forall c \in C_{CV}, k \in K_{CV}, i \in V_0, j \in V_{n+1} \tag{7.39}$$

Balance of load flow at each node is defined through constraints 7.40 and 7.41. These constraints model the vehicle load flow as increasing by the amount of cargo demand of each visited demand node. Constraints 7.42–7.45 link the vehicle load on each arc with the corresponding load interval $m$ so that the proper load interval is used while estimating the vehicle energy requirements along each arc.

$$\sum_{j \in V_0'} U_{jick}^{EV} - \sum_{j \in V_{N+1}'} U_{ijck}^{EV} = q_i \sum_{j \in V_{N+1}'} \sum_t \sum_m x_{ijcktm}^{EV} \quad \forall c \in C_{EV}, k \in K_{EV}, i \in V$$

$$\tag{7.40}$$

$$\sum_{j \in V_0} U_{jick}^{CV} - \sum_{j \in V_{N+1}} U_{ijck}^{CV} = q_i \sum_{j \in V_{N+1}} \sum_t \sum_m x_{ijcktm}^{CV} \quad \forall c \in C_{CV}, k \in K_{CV}, i \in V$$

$$\tag{7.41}$$

$$U_{jick}^{CV} \ge LU_m \sum_t x_{ijcktm}^{CV} \quad \forall c \in C_{CV}, k \in K_{CV}, i \in V_0, j \in V_{n+1}, m \in M$$

$$\tag{7.42}$$

$$U_{jick}^{CV} - UU_m \le M \left(1 - \sum_t x_{ijcktm}^{CV}\right)$$

$$\forall c \in C_{CV}, k \in K_{CV}, i \in V_0, j \in V_{n+1}, m \in M \tag{7.43}$$

$$U_{jick}^{EV} \ge LU_m \sum_t x_{ijcktm}^{EV} \quad \forall c \in C_{EV}, k \in K_{EV}, i \in V_0', j \in V_{n+1}', m \in M$$

$$\tag{7.44}$$

$$U_{jick}^{EV} - UU_m \le M \left(1 - \sum_t x_{ijcktm}^{EV}\right)$$

$$\forall c \in C_{EV}, k \in K_{EV}, i \in V_0', j \in V_{n+1}', m \in M \tag{7.45}$$

In constraint 7.46, the remaining battery capacities of all electric vehicles are set to their full battery capacity before starting their route. This means that all the vehicles leave the depot with fully charged batteries. Constraint 7.47 sets the battery level of a vehicle arriving at a node succeeding a demand node in accordance with the energy consumption on the arc joining these two nodes. Constraint 7.48 defines the same relation for the nodes succeeding a charging station.

$$R_{0ck} = BC_c^{EV} \qquad \forall c \in C_{EV}, k \in K_{EV} \qquad (7.46)$$

$$R_{jck} \leq R_{ick} - \left( \alpha'_{ij} \times w_c^{EV} + \beta'_{ij} \times \left( intercept_{ij}^t \times \sum_m x_{ijcktm}^{EV} + slope_{ij}^t \right. \right.$$

$$\times \sum_m \left( DT_{ijcktm}^{EV} - LT_t \times x_{ijcktm}^{EV} \right) \right) + \alpha'_{ij}$$

$$\times \left( intercept_{ij}^t \times \sum_m M_{avg} \times x_{ijcktm}^{EV} + slope_{ij}^t \right.$$

$$\left. \times \sum_m \overline{U}_m \times \left( DT_{ijcktm}^{EV} - LT_t \times x_{ijcktm}^{EV} \right) \right) + EC_k \left( 1 - \sum_m x_{ijcktm}^{EV} \right) \right)$$

$$\forall c \in C_{EV}, k \in K_{EV}, i \in V, j \in V'_{n+1}, t \in T, i \neq j \qquad (7.47)$$

$$R_{jck} \leq EC_k - \left( \alpha'_{ij} \times w_c^{EV} + \beta'_{ij} \times \left( intercept_{ij}^t \times \sum_m x_{ijcktm}^{EV} + slope_{ij}^t \right. \right.$$

$$\times \sum_m \left( DT_{ijcktm}^{EV} - LT_t \times x_{ijcktm}^{EV} \right) \right) + \alpha'_{ij}$$

$$\times \left( intercept_{ij}^t \times \sum_m M_{avg} \times x_{ijcktm}^{EV} + slope_{ij}^t \right.$$

$$\left. \times \sum_m \overline{U}_m \times \left( DT_{ijcktm}^{EV} - LT_t \times x_{ijcktm}^{EV} \right) \right) \right)$$

$$\forall c \in C_{EV}, k \in K_{EV}, i \in F' \cup 0, j \in V'_{n+1}, i \neq j, t \in T \qquad (7.48)$$

Constraint 7.49 ensures that if an internal combustion engine vehicle visits any demand node in a Low Emission Zone, the decision variable, $y_{ckl}^{CV}$, is set to one for the vehicle and the corresponding zone. Finally, binary decision variables are defined in constraints 7.50.

$$\sum_{i \in l} \sum_{j \in V_{N+1}} \sum_t \sum_m x_{ijcktm}^{CV} \leq 100 y_{ckl}^{CV} \qquad \forall c \in C_{CV}, k \in K_{CV}, l \in LEZ_l$$

$$\qquad (7.49)$$

$$x_{ijcktm}^{EV}, x_{ijcktm}^{CV}, y_{ckl}^{CV} \in 0, 1 \qquad (7.50)$$

## 7.3.2  Numerical Study

In order to verify the performance of developed mathematical model in finding optimal solution for the introduced variant of green vehicle routing problem, it was used to solve a GVRP on a small size network for two different scenarios: (a) where travel time on arcs is constant over the whole planning period and (b) a GVRP with time-dependent travel times. Both problems were defined on a similar network with the same demand distribution and one LEZ, shown by dotted lines in Fig. 7.3. The problem was solved for the two travel time scenarios using the Xpress commercial solver.

The solution found by Xpress is represented in Fig. 7.3. In this figure, the green and black lines represent ECV and ICCV routes, respectively. While for both scenarios, the optimal fleet size was found to be the same, the routing of the vehicles was different between the two scenarios. As it can be seen, the optimal route for the ECV is found to be the shortest distance route (Depot $\to P_3 \to P_5 \to CS \to P_6$), if the travel time on arcs is assumed to be constant during the operation hours. On the other hand, for the time-dependent travel time scenario, the sequence of the visit to demand nodes and charging station is changed to accommodate the change in energy requirement due to the different levels of congestion on routes during different times of a day. Therefore, it can be concluded that the developed mathematical model can take into account variations in travel time as well as limitations is adoption of the two types of vehicle.

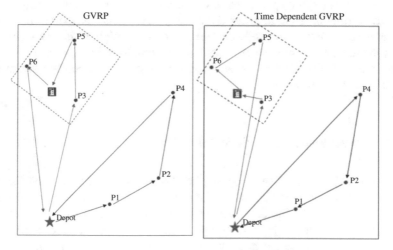

**Fig. 7.3**  Minimum cost routes—static vs. time-dependent GVRP

## 7.4   Heuristic Method

It is well known that VRP is an NP-hard problem [22], and as the size of the network increases, the computational complexity of the problem grows exponentially and commercial solvers become unable to find optimal solutions to the problem. Therefore, it is essential to develop an efficient heuristic algorithm that can find sound solutions to the problem in a reasonable amount of time. The heuristic algorithm developed in this study is based on Ruin and Recreate (RR) approach, a new class of algorithms introduced by Schrimpf et al. [27] used to solve VRP with time window instances. The basic idea behind this algorithm is to obtain new solutions by deconstructing an existing feasible solution and rebuilding it by following a set of procedures. First, an initial feasible solution is generated using a constructive heuristic approach. Then the generated solution is improved by deconstructing and rebuilding new solutions from the current solution until some stopping criteria were met.

The proposed heuristic algorithm shares some similarities with the classic local search (LS) approach, but it presents some advantages that lead to feasible solutions with better quality. While both of the methods make use of a systematic perturbation of the current solution that leads to a new solution, they are based on different concepts of a neighborhood. In the LS approach, a new solution is achieved by a small modification to the current solution, such as the movement of a customer to a different route. In fact, LS performs a deep evaluation of solutions close to the current one and chooses a better neighboring solution. However, the algorithm proposed in this research generates new solution by deconstructing a larger part of the current solution. Therefore, our algorithm not only relies on exploring solutions close to the current one but also evaluates solutions that might be far from the current one in the feasible solution space. Looking into solutions far from the current one, which is called diversification strategy, is rarely applied during the evolution of search in the LS algorithms.

### 7.4.1   Constructive Heuristic Algorithm

The constructive phase of the proposed heuristics is designed based on the regret procedure presented in studies by Christofides et al.[8] and Potvin and Rousseau [24]. Figure 7.4 shows the flow chart of the constructive heuristic. The algorithm starts with generating a partial solution comprised of a set of feasible single-customer routes. To generate the initial partial solution, the network plane is divided into $n$ identical cones with origins at the depot, where $n$ is calculated as the sum of customer demands multiplied by 1.5, divided by the maximum capacity of available vehicles. Then, in each cone, a single-customer route is defined that serves the demand node farthest from the depot with the largest demand size.

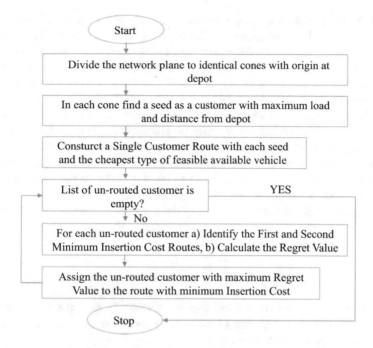

**Fig. 7.4** Constructive heuristic workflow

The generated partial solution is completed by iteratively assigning each unrouted demand point either to the existing routes or to a new one. The unrouted customers are chosen and assigned to the routes based on a regret score, $\delta(i)$, accounting for the urgency of assigning customer $i$ to a route. The computation of $\delta(i)$ is based on a penalty, $\rho_r(i)$, associated with the cost of assigning customer $i$ to route $r$. This penalty is comprised of six components and is calculated using Eq. 7.51. In this equation, $Set_r(i)$ denotes the set of all possible insertion points for customer $i$ in route $r$. In fact, the penalty of inserting customer $i$ in route $r$ is calculated for (a) all possible insertion points in the $Set_r(i)$ and (b) for the two types of vehicles, ICCV and ECV, then, the minimum penalty is assigned to $\rho_r(i)$.

$$\rho_r(i) = min_{k \in \{ICCV, ECV\}} \Big\{ min_{p \in Set_r(i)} \Big\{ extra - energy - req_r(i, p, k)$$

$$+ veh - chg - cost_r(i, k) + time - window - pen_r(i, p, k)$$

$$+ LEZ - pen_r(i, p, k) + bat - deg_r(i, p, k) + carbon - permit(i, p, k)$$

$$+ extra - Labor - Cost(i, p, k) \Big\} \Big\} \tag{7.51}$$

The first component of Eq. 7.51 is the extra energy required by the vehicle to travel to and from the node after its insertion in the route. The second component is the vehicle change cost if the vehicle serving route $r$ must be changed to visit customer $i$. The third component is the time window penalty if the insertion of customer $i$ to the route $r$ results in service time window violations at any demand nodes. The fourth component is the LEZ penalty cost that incurs when customer $i$ is in LEZ and route $r$ is served by an ICCV. The fourth component is the battery degradation cost, which is applicable only if customer $i$ is inserted in a route served by an electric truck. If the extra energy requirements of the electric truck to serve customer $i$ result in the battery capacity violations, there is a need for the battery to be exchanged with a full one and battery degradation cost should be accounted for. The fifth component is the carbon emission cost, which is only applicable in the case of ICCVs. And the last one is the extra labor cost due to the increase in travel time after the insertion of the new customer to the route.

Having estimated all the components of the penalty associated with the insertion of each unrouted customer $i$ to the existing routes in the partial solution, the regret score $\delta(i)$ is estimated as:

$$\delta(i) = smin_r(\rho_r(i)) - min_r(\rho_r(i)) \qquad (7.52)$$

where $smin_r$ denotes the second minimum value. In fact, the score $\delta(i)$ is the difference between the penalties of the second minimum cost insertion route and the first minimum cost insertion route.

Having estimated the value of score $\delta(i)$ for all unrouted customers, the one with the largest $\delta$ is selected and assigned to the minimum insertion cost route. Once a customer is assigned to a route $r$, the $\delta$ scores are updated for the set of unrouted customers, and this process is repeated until all customers are routed. The pseudocode for the proposed constructive heuristic is given in Algorithm 2.

---

**Algorithm 2:** Constructive heuristic algorithm pseudocode

---
1      **Input**:Partial Solution]
2      **Output**: Initial Feasible Solution
       **for** unrouted customer $i$ **do**
           **for** each route $r$ **do**
               compute the score $\rho_r(i)$;
               Compute the score $\delta(i)$;
           **end for**
       **end for**
       **while** the set of unrouted customers is not empty **do**
           Identify the customer with the maximum value of $\delta(i)$;
           Insert the customer $i$ in the best possible insertion point of the route $r$ with
               minimum $\rho_r(i)$;
           **for** each unrouted customer $i$ **do**
               update the scores $\rho_r(i)$ and $\delta(i)$;
           **end for**
       **end while**

---

## 7.4.2   Improvement Heuristic Algorithm

As mentioned before, the heuristic algorithm proposed in this study is based on the deconstruction of a large part of the current solution and rebuilding it to generate better new solutions. Therefore, once a feasible initial solution is generated using the constructive heuristic, it is destroyed based on a strategy and is rebuilt using the same procedure explained in the constructive heuristic. Figure 7.5 shows the flow chart of the improvement heuristic. To destruct the current solution, first, a target route is identified. The target route is defined as a route that could be served by a cheaper vehicle if a small set of customers assigned to it was removed. In other words, the target route is a route minimizing excess load, which is defined as the nonnegative quantity of goods that could not be delivered to the customers served by route $r$ using a cheaper vehicle of type $h - 1$, and it is calculated using the equation below:

$$EL_r = TL_r - Q_{h-1} \tag{7.53}$$

where $EL_r$ is the excess load of route $r$, $TL_r$ is total load carried by vehicle serving route $r$, and $Q_{h-1}$ is the loading capacity of the cheaper vehicle of type $h - 1$.

Once a route is selected as a target route, it is stored in a tabu list with an infinite length such that it cannot be selected in any other iteration during the entire execution of the algorithm. When there are no more target routes found by

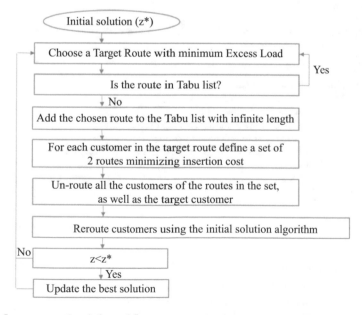

**Fig. 7.5** Improvement heuristic workflow

the heuristic, the algorithm is prematurely terminated, and the current solution is accepted as the best solution found by the algorithm. Therefore, one of the stopping criteria in the improvement phase is the failure to find a target route.

Given a target route, $r$, each of its customers is considered as a target customer used to initialize the set of customers to be removed from the current solution. For each target customer in the target route, two routes with minimum insertion cost are found for a target customer, and all the customers in those two routes are unrouted as well as the target customer. Then, the resulting partial solution is transformed into a complete solution by using the regret algorithm used in the constructive heuristic. In fact, the unrouted customers are added to existing or new routes iteratively by using regret scores until no customer is left and the solution is complete. If the new solution is better than the previous solution, it is accepted as the new current solution, and it is used in the next iteration of the improvement phase.

On the other hand, if the new solution is not better than the previous solution, the previous solution remains as the best current solution for the next iteration of the improvement phase. This procedure is continued until a time limit was reached or the algorithm fails to find a target route. The pseudocode for the improvement heuristic is given in Algorithm 3.

---

**Algorithm 3:** Improvement heuristic algorithm pseudocode

| | |
|---|---|
| 1 | **Input**: initial feasible solution to TDGVRP instance $[l_0, z_0]$ |
| 2 | **Output**: $[l^*, z^*]$ |
| 3 | **Initialization**: $l^* = l_0, z^* = z_0$ |
| | **while** not time limit **do** |
| | $l = l^*$ |
| | Determine the non-tabu route $\bar{r}$ minimizing the excess load; |
| | **if** no such route exists **then** |
| | stop; |
| | **end if** |
| | Store $\bar{r}$ in tabu list with infinite length; |
| | **for** each target customer $i$ in target route $\bar{r}$ **do** |
| | Remove customer $i$ from route $\bar{r}$ of $l$; |
| | Find the set of 2 routes with minimum insertion cost for $i$; |
| | Remove all customers of the 2 routes from $l$; |
| | Use the constructive heuristic to generate a new feasible solution $z$; |
| | **if** $z < z^*$ **then** |
| | $z^* = z, l^* = l$ |
| | **end if** |
| | **end for** |
| | **end while** |

---

The evolution of steps in the improvement phase for a network of 10 customers is shown in Figs. 7.6, 7.7, 7.8, 7.9, and 7.10. In this network, there is one LEZ that includes three demand nodes and one charging station. The numbers in the boxes show the total load carried by the vehicles assigned to the routes, and the optimal fleet size is listed in the table next to graphs. The routes in Fig. 7.6 form the initial

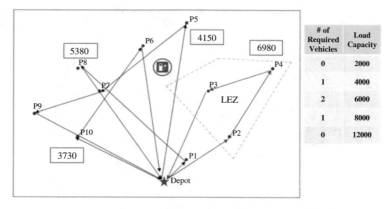

**Fig. 7.6** Evolution of the improvement heuristic algorithm—Part a

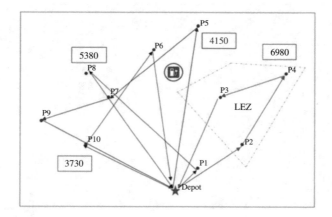

**Fig. 7.7** Evolution of the improvement heuristic algorithm—Part b

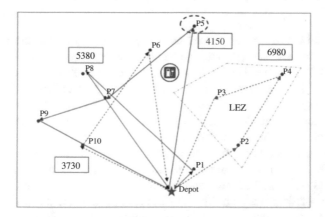

**Fig. 7.8** Evolution of the improvement heuristic algorithm—Part c

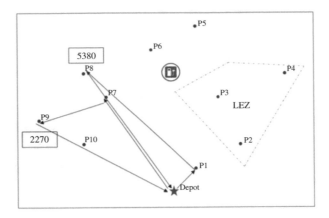

**Fig. 7.9** Evolution of the improvement heuristic algorithm—Part d

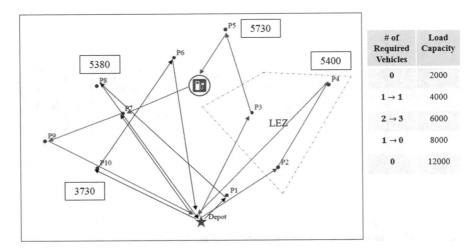

**Fig. 7.10** Evolution of the improvement heuristic algorithm—Part e

solution found by constructive heuristic. Based on this solution, the optimal fleet size required to serve the customer demands in the network is one ICCV with the capacity of 4000 lbs, two ICCVs with the capacity of 6000 lbs, and one ECV with the capacity of 8000 lbs. The target route minimizing the excess load is shown in the red color in Fig. 7.7. As mentioned before, once the target route is identified, each of its customers is considered as a target customer, and for each target customer, two routes with the minimum insertion cost are identified. In Fig. 7.8, demand node P5 is chosen as a target customer, and the two routes with minimum insertion cost for this node are shown in dotted lines. In Fig. 7.9, all the customers in the two identified routes, as well as the target customer, are unrouted, resulting in a partial solution. The partial solution is rebuilt by the constructive heuristic in Fig. 7.10, and new routes are generated. As it can be seen in this figure, the fleet size has changed to one ICCV with the capacity of 4000 lbs, and two ICCVs and one ECV with the battery capacity of 6000 lbs. In fact, the total fleet cost in the improvement phase is decreased by changing the fleet composition more efficiently.


## 7.4.3   Verification of the Heuristic Algorithm for Time-Dependent GVRP

To verify the efficiency of the proposed heuristic algorithm in finding sound solutions, its performance was verified by comparing the solutions generated by the algorithm with optimal solutions found by Xpress commercial solver for small size problems. For this purpose, a set of small size problems was defined. For each problem, a random network of demand nodes, charging stations, and depot were generated. The networks had different topologies of customers, charging station, and depot locations. For all the problem instances, it was assumed that 2 types of electric trucks, with different loading and battery capacities, and 2 types of internal combustion engine trucks, with different loading capacities, are available. The planning period was divided into three time windows with different travel time functions defined for each time period. Because rush hour traffic might be different for opposite directions of a route, different travel time functions were defined for the two opposite directions of an arc in the network. Average acceleration rate on each direction of an arc in the network was randomly generated from a uniform distribution for each time period. The acceleration rates during morning and evening rush hours were randomly generated from the range $[0.3, 0.5]$ m/s$^2$. For the midday off-peak, the acceleration rate on arcs was randomly generated from the range $[0.1, 0.3]$ m/s$^2$. The customer demands were randomly generated from uniform distributions listed in Table 7.2. In all of the cases, it was assumed that all operated vehicles leave the depot at 8 am and all the ECVs are fully charged overnight.

Each one of the defined problems was solved with both Xpress commercial solver and the proposed heuristic method coded in Python, and the results were compared. The heuristic running time, Xpress running time, and solution gap are shown for each problem in Table 7.2. As it can be seen in this table, the results show that the proposed heuristic algorithm works very well. While it might take more than 2 days for the Xpress commercial solver to find solutions to the small size problem, the proposed heuristic algorithm is capable of finding sound solutions with an average gap of 2.9% in a matter of seconds.


## 7.4.4   Case Study

To evaluate the efficiency of the developed heuristic algorithm in finding sound solutions to real-world sized problems, it was tested on a large size network. Since access to data on real-world delivery operations such as FedEx or UPS was not

**Table 7.2** Heuristic solution vs. optimal solution

| No. of demand nodes/CS | No. of nodes in LEZ/LEZ cost | Carbon price ($/gram) | Emission cap (gram/day) | No. of constraints/variables | Xpress running time (s) | Heuristic running time (s) | Solution gap |
|---|---|---|---|---|---|---|---|
| 5/1 | 1/100 | 0.5 | 30 | 11,844/8227 | 3004 | 2.2 | 0.4% |
| 5/1 | 1/100 | 1 | 30 | 11,844/8227 | 4108 | 1.8 | 0.1% |
| 5/1 | 1/20 | 0.5 | 20 | 11,844/8227 | 10,723 | 1.7 | 2.8% |
| 5/1 | 1/20 | 1 | 20 | 11,844/8227 | 7283 | 2.1 | 1.5% |
| 6/1 | 2/20 | 0.5 | 30 | 16,577/10,967 | >2 Day | 2.3 | 7.6% |
| 6/1 | 2/20 | 1 | 30 | 16,577/10,967 | 27,220 | 1.8 | 2.2% |
| 6/1 | 2/100 | 2 | 20 | 16,577/10,967 | 21,720 | 1.9 | 0.3% |
| 6/1 | 2/100 | 2 | 20 | 16,577/10,967 | 25,274 | 1.7 | 1.8% |
| 6/1 | 2/20 | 1.5 | 20 | 16,577/10,967 | >2 Day | 1.9 | 9.8% |
| 6/1 | 2/30 | 1 | 20 | 16,577/10,967 | 20,836 | 2.1 | 2.3% |

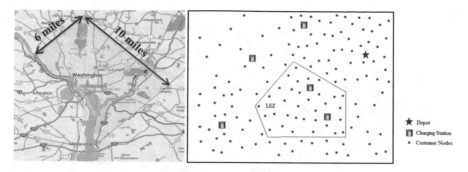

**Fig. 7.11** Randomly generated network for case study

available for this study, a network of 150 demand nodes was randomly generated in an area with the size of Washington DC, as shown in Fig. 7.11. The generated network consisted of 150 demand nodes, one depot, and five charging stations.

It was assumed all vehicles depart from the depot at 8 in the morning and return to the depot at 6 in the evening. Therefore, three different travel time windows were considered for the operation: morning rush hour, midday off-peak, and evening rush hour. To account for the variation of congestion level during the defined time windows, three different travel time functions were considered for each arc of the network along each direction.

Moreover, an average acceleration rate was randomly generated for each direction of arcs in the network during each time window. The quantity of demand and service time window for each customer in the network was randomly generated from uniform distributions. A Low Emission Zone with the penalty of $60 for ICCVs was defined on the network. An emission cap of 200 grams/day was considered on the emissions produced by delivery operations. The price of a carbon permit was set to $0.50/gram. Therefore, any extra permit was purchased or sold at this price. The price of fuel and electricity was set to $0.76/gram and $0.12/kwh, respectively. The value of time was assumed to be $15/h. It was assumed three different types of ICCVs and ECVs are available with different loading and battery capacities.

The proposed heuristic method was used to find a solution to the defined problem. Figure 7.12 illustrates the improvement in the heuristic solution as the running time increases. The best solution is achieved by running the algorithm for 4300 s, and no significant improvements is achieved by running the heuristic for longer durations. If not stopped at 4300 s, the algorithm would run until there is no target route found which in this case was at 8000 s.

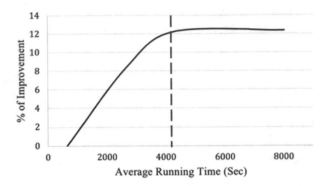

**Fig. 7.12** Improvement in heuristic solution over running time

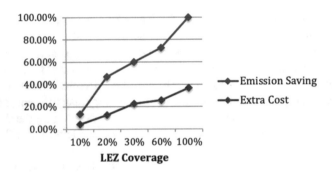

**Fig. 7.13** Emission saving and extra cost over different LEZ coverages

## 7.4.5 Sensitivity Analysis

There are different parameters in the problem that could have a significant effect on the solution found by the heuristic. To evaluate the effect of these parameters on the solution, a set of sensitivity analysis is performed. First, a time-dependent GVRP is solved for a basic scenario, defined on the randomly generated network shown in Fig. 7.11, where there are no LEZ or emission cap regulations present. Then different parameters in the problem are changed, and the variations in the solution are investigated.

Figure 7.13 illustrates the change in operation cost versus the change in total emission as the LEZ coverage increases. As it is expected, by increasing the LEZ coverage, a greater number of ICCVs are replaced by ECVs and, as a result, the total emission produced by vehicles while routing is reduced. The minimum LEZ coverage of 10% is seen to reduce the emission by 13% while increasing the cost of operation by only 4%. Obviously, the maximum reduction in emission is achieved when all demand nodes are located within a LEZ. Replacing all ICCVs with ECVs results in almost 37% increase in the operation cost.

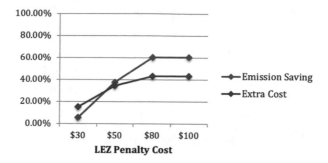

**Fig. 7.14** Emission saving and extra cost over different LEZ penalty costs

Another factor that affects the fleet mix and route of the vehicles is the LEZ penalty cost. If the operation cost of ICCV plus the LEZ penalty cost is more than the operation cost of ECV, then an electric truck replaces the conventional truck. Therefore, it is expected to have different fleet mix and routing plans with different LEZ penalty costs. To verify the sensitivity of the heuristic to LEZ penalty cost, four different scenarios were considered with the same level of LEZ coverage but different penalty costs. Figure 7.14 shows the changes in the operation cost and emission savings as LEZ penalty changes. As can be seen, when there is a LEZ with the penalty cost of $30, some of the ICCVs are replaced with ECV. This change in the fleet size increases the operation cost and slightly reduces the routing emissions. However, when the penalty cost increases from $30 to $50, more number of the ICCVs are replaced with ECVs, resulting in a more significant reduction in emissions. This change in the fleet size continues when the LEZ penalty cost increases to $80, but it stops afterward meaning that no changes would be expected in the fleet size and routing plan of vehicles for higher penalty cost. This is because, with the penalty cost of $80, all the demand nodes in the LEZ are served by ECVs and increasing the penalty cost to higher values does not affect the fleet mix.

Another factor affecting the fleet size and routing plan of trucks is the emission cap and trade policy. If there is a cap imposed on the acceptable level of emission, there might be a change in fleet size and routing plans. Extend of this change depends on the penalty charged for extra emission known as carbon permit cost. Figure 7.15 shows the changes in the solution as a result of changes in emission price and cap. As the price of carbon increases from $1 to $5, more ICCVs are replaced with ECVs until the fleet becomes pure ECVs. While there is an overall increasing trend in emission savings over different carbon prices, the change in total operation cost does not follow a constant trend. It first increases and then starts to decrease. The decrease in total cost happens when the benefits earned by selling extra carbon permits outweigh the extra employment cost of ECVs. It should be noted that the solutions to the cases are the same for both emission caps in terms of fleet mix and routing cost, but the total operation cost is different. It is expected due to the different levels of benefit achieved by selling extra carbon permit cost.

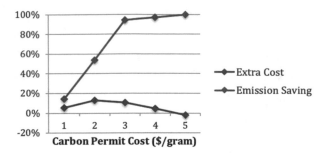

**Fig. 7.15** Emission saving and extra cost over different emission costs—emission cap = 200

## 7.5 Summary, Conclusion, and Recommendations for Future Research

In this research, a special case of time-dependent GVRP with a mixed fleet of heterogeneous electric and internal combustion engine commercial vehicles was studied. A mathematical formulation was developed to find optimal solutions to the defined problem accounting for variations in travel time during different time periods of a day. Through the developed formulation, the optimal fleet size and the minimum cost routes were found. The cost included the vehicle fixed cost, as well as the routing cost in the form of electric energy requirement or fuel consumption, labor cost, customer inconvenience costs that result from breaking service time windows, Low Emission Zone penalty cost, and carbon permit cost.

While previous studies in the field of GVRP with a mixed fleet of electric and conventional vehicles focus only on the limitations of electric vehicles such as high purchase cost and limited driving range, the variant of GVRP studied in this research was defined to account for the limitations of both vehicle types. These limitations were in the form of Low Emission Zone penalty cost and emission cap for internal combustion engine vehicles and high purchase cost and limited driving range for electric vehicles. Low Emission Zone and emission cap are the policies used by the government in some countries to encourage the use of green vehicles by imposing some limitations on the use of conventional vehicles. Therefore, the solution to the time-dependent GVRP defined in this study finds the best fleet size and routing plan to serve customers by taking into account the trade-offs between the two types of vehicles. Moreover, taking into account the time dependency of travel times enabled more accurate estimation of vehicle energy requirements, which is even more important in the case of electric vehicle due to the high sensitivity of electric vehicle's driving range to energy consumption rate. To account for the variations in travel time, continuous travel time functions were used. The use of continuous travel time functions instead of step functions assured the conservancy of the FIFO concept and provided a more realistic presentation of travel time variations.

A heuristic algorithm was proposed to solve the problem for large size networks. The heuristic was defined based on the Ruin and Recreate (RR) approach introduced by Schrimpf, et al. [27]. This algorithm was preferred over the local search algorithms as it is expected to provide better solutions due to the diversification effect embedded in it by deconstructing a large part of the solution. To verify the performance of the proposed heuristic method, the results from the heuristic were compared to the solutions found by Xpress commercial solver. The comparison between the Xpress and heuristic solutions showed that the proposed heuristic method is capable of producing good results within a very short time.

To verify the performance of the heuristic on real-world large size problems, the proposed heuristic method was used to solve a case study defined on a randomly generated network with the size of Washington DC. Moreover, to see how the heuristic method is performing when the parameters in the problem are changing, an extensive sensitivity analysis was performed on parameters such as LEZ coverage, LEZ penalty cost, and emission price.

Overall, it was concluded that the proposed formulation for time-dependent GVRP could handle trade-offs between the limitations on both types of vehicles, as well as variations in travel time during different periods of operation. The proposed heuristic method was found to perform well and produced fairly good results for the generated test problems in terms of solution accuracy and run time when compared with the exact solution for small problems. The result of the sensitivity analysis performed on the solution of heuristic with respect to parameters of LEZ coverage, LEZ penalty cost, and emission cap showed the effectiveness of the heuristic algorithm in considering the trade-offs between ICCVs and ECVs under different LEZs and cap and trade policies.

Although it was shown that the developed heuristic algorithm can be used to identify changes in fleet design and routing of last-mile delivery operations as a result of green logistics policies, there are several interesting avenues for future research. In this research, the heuristic solutions were compared with the Xpress optimal solutions. This comparison showed good results and the errors were acceptable in the examples compared. However, the comparison was made only on the examples that Xpress could handle. Xpress cannot solve large size problems even in a very long running time, so the comparison of the heuristic method with optimal solutions is not made on the real size problems. The systematic approach to see how the heuristic method performs is to find a good lower bound for this problem that is missing in this research and is highly recommended for future studies.

Moreover, the problem studied in this research was defined on a network with a single depot. However, in real-world operations, there might be more than one depot serving demand in the network. Charging stations are assumed to be of battery swapping type with constant service time for all vehicles regardless of the level of the battery available upon a visit to a charging station. Although battery swapping stations are currently in practice and might be the future of charging stations, developing a model with varying charging times based on the battery level upon visit could enhance the solutions found by the model developed in this research.

# References

1. Alwakiel, H.N.: Leveraging weigh-in-motion (WIM) data to estimate link-based heavy-vehicle emissions. PhD Thesis. Portland State University (2011)
2. Ardekani, S., Hauer, E., Jamei, B.: Traffic impact models. In: Traffic Flow Theory, pp. 1–7. National Research Council, Washington (1996)
3. Barco, J., Guerra, A., Muoz, L., Quijano, N.: Optimal Routing and Scheduling of Charge for Electric Vehicles: Case Study. Universidad de los andes, Bogota (2012). http://arxiv.org/ftp/arxiv/papers/1310/1310.0145.pdf
4. Bard, J., Huang, L., Dror, M., Jaillet, P.: A branch and cut algorithm for the VRP with satellite facilities. IIE Trans. **30**, 821–834 (1998)
5. Bektas, T., Laporte, G.: The Pollution-Routing problem. Transp. Res. Part B Methodol. **45**(8), 1232–1250 (2011)
6. Bigazzi, A.Y., Bertini, R.L.: Adding green performance metrics to a transportation data archive. Transp. Res. Rec. J. Transp. Res. Board **2121**(1), 30–40 (2009)
7. Bruglieri, M., Pezzella, F., Pisacane, O., Suraci, S.: A variable neighborhood search branching for the electric vehicle routing problem with time windows. Electr. Notes Discrete Math. **47**, 221–228 (2015)
8. Christofides, N., Mingozzi, A., Toth, P.: The vehicle routing problem. In: Combinatorial Optimization, pp. 315–338. Wiley, Chichester (1979)
9. Davis, B.A., Figliozzi, M.A.: A methodology to evaluate the competitiveness of electric delivery trucks. Transp. Res. Part E Logist. Transp. Rev. **49**(1) 8–23 (2013)
10. Demir, E., Bektas, T., Laporte, G.: A comparative analysis of several vehicle emission models for road freight transportation. Transp. Res. Part D Trans. Envir. **6**(5), 347–357 (2011)
11. Demir, E., Bektas, T., Laporte, G.: The bi-objective pollution-routing problem. Eur. J. Oper. Res. **232**(3), 464–478 (2013)
12. Desaulniers, G., Errico, F., Irnich, S., Schneider, M.: Exact algorithms for electric vehicle-routing problems with time windows. Les Cahiers du GERAD G-2014-110, GERAD, Montréal (2014)
13. Eglese, R.W., Black, D.: Optimizing the routing of vehicles. In: McKinnon, A., Cullinane, S., Browne, M., Whiteing, W. (eds.) Green Logistics: Improving the Environmental Sustainability of Logistics, pp. 215–228. KoganPage, Great Britain (2010)
14. Erdoğan, S., E. Miller-Hooks: A green vehicle routing problem. Trans. Res. Part E Log. Trans. Rev. **48**(1), 100–114 (2012)
15. Felipe, À., Ortuño , M.T., Righini, G., Tirado, G.: A heuristic approach for the green vehicle routing problem with multiple technologies and partial recharges. Transp. Res. Part E Logist. Transp. Rev. **71**, 111–128 (2014)
16. Goeke, D., Schneider, M.: Routing a mixed fleet of electric and conventional vehicles. Eur. J. Oper. Res. **245**(1),81–99 (2015)
17. Hiermann, G., Puchinger, J., Hartl, R.F.: The electric fleet size and mix vehicle routing problem with time windows and recharging stations. Working Paper. http://prolog.univie.ac.at/research/publications/downloads/$Hie_2014_638$.pdf (2014). Accessed November 2015
18. Jovicic, N.M., Boškovic, G.B., Vujic G.V. et al: Route optimization to increase energy efficiency and reduce fuel consumption of communal vehicles Thermal Science **14**, 67–78 (2011)
19. Kara, I., Kara, B.Y., Yetis, M.K.: Energy minimizing vehicle routing problem. In: Dress A., Xu Y., Zhu B. (eds.) Combinatorial Optimization and Applications. Lecture Notes in Computer Science vol. 4616. Springer, Berlin (2007)
20. Kuo, Y.: Using simulated annealing to minimize fuel consumption for the time-dependent vehicle routing problem. Comput. Ind. Eng. **59**(1), 157–165 (2010)
21. Laporte, G., Nobert, Y., Desrochers, M.: Optimal routing under capacity and distance restrictions. Oper. Res. **33**(5), 1050–1073 (1985)

22. Lenstra, J.K., Kan, A.H.: Complexity of vehicle routing and scheduling problems. Networks **11**, 221–227 (1981)
23. Maden, W., Eglese, R.W., Black, D.: Vehicle routing and scheduling with time-varying data: a case study. J. Oper. Res. Soc. **61**(3), 515–522 (2010)
24. Potvin, J.Y., Rousseau, J.M.: A parallel route building algorithm for the vehicle routing scheduling problem with time windows. Eur. J. Oper. Res. **66**, 331–340 (1993)
25. Sassi, O., Cherif, W.R., Oulamara, A.: Vehicle routing problem with mixed fleet of conventional and heterogenous electric vehicles and time dependent charging costs. https://hal.archives-ouvertes.fr/hal-01083966/ (2014). Accessed November 2015
26. Schneider, M., Stenger, A., Goeke, D.: The electric vehicle routing problem with time windows and recharging stations. Trans. Sci. **48**, 500–520 (2014)
27. Schrimpf, G., Schneider, J., Stamm-Wilmbrandt, H., Dueck, G.: Record breaking optimization results using the ruin and recreate principle. J. Comput. Phys. **159**, 139–171 (2000)
28. Taefi T., Kreutzfeldt, J., Held, T., Konings, R., Kotter, R., Lilley, S., Baster, H., Green, N., Stie Laugesen, M., Jacobsson, S., Borgqvist, M., Nyquist, C.: Comparative analysis of European examples of freight electric vehicles scheme. A systematic case study approach with examples from Denmark, Germany, the Netherlands, Sweden, and the UK. In: 4th International Conference on Dynamic in Logistics (LDIC 2014). Bremen (2014)
29. Tavares, G., Zsigraiova, Z., Semiao, V., da Graça Carvalho, M.: A case study of fuel savings through optimisation of MSW transportation routes. Manag. Envir. Qual. **19**(4), 444–454 (2008)
30. Tavares, G., Zsigraiova, Z., Semiao, V., Carvalho, M.G.: Optimisation of MSW collection routes for minimum fuel consumption using 3D GIS modelling. Waste Manag. **29**(3), 1176–1185 (2009)
31. US DOT, Department of Transportation: The Transportation's Role in Reducing US Greenhouse Gas Emissions (2010)
32. Xiao, Y., Zhao, Q., Kaku, I., Xu, Y.: Development of a fuel consumption optimization model for the capacitated vehicle routing problem. Comput. Oper. Res. **39**(7), 1419–1431 (2012)

# Chapter 8
# Recent Developments in Real Life Vehicle Routing Problem Applications

**Lina Simeonova, Niaz Wassan, Naveed Wassan, and Said Salhi**

**Abstract** In recent years, there is an increasing volume of research focusing on problems with real life applicability across the entire vehicle routing domain. Bringing academia closer to industry operations has enriched the literature significantly and led to the creation of multiple vehicle routing problem variants and streams. This chapter discusses some of the main vehicle routing problem categories motivated by real operations, namely integrated problems, problems with alternative objective functions and those problems which are highly problem specific and have unique characteristics. We provide a discussion on the evolution of real life routing problems and evaluate some important methodological considerations. We also discuss aspects, which are typically overlooked in the literature such as the role of data, methodological comparability, implementation and benefits realization. Finally, we offer guidance for best practice and future research in the area of real life routing problems.

## 8.1 Introduction

The evolution of Vehicle Routing Problem (VRP) and its variants is inspired by real life operations and there is a noticeable trend in the literature to bring VRP research closer to real life routing practice. The literature on real life vehicle routing problems (RVRPs) has developed rapidly since 2006, motivated by the different supply chain and routing challenges faced by businesses on daily basis. Academics increasingly

L. Simeonova (✉) · S. Salhi
Centre for Logistics and Heuristic Optimisation (CLHO), Kent Business School, University of Kent, Canterbury, UK
e-mail: L.Simeonova@kent.ac.uk; S.Salhi@kent.ac.uk

N. Wassan
Department of Operations Management and Business Statistics, Suntan Qaboos University, Muscat, Oman

Sukkur IBA University, Sukkur, Pakistan
e-mail: n.wassan@squ.edu.om; naveed.wassan@iba-suk.edu.pk

© Springer Nature Switzerland AG 2020
H. Derbel et al. (eds.), *Green Transportation and New Advances in Vehicle Routing Problems*, https://doi.org/10.1007/978-3-030-45312-1_8

work with practitioners to understand the important aspects of routing and some of the main trends in Operations Research (OR) have been influenced by bridging the gap between industry and academia. Accommodating various real life attributes has enriched the literature of the VRP variants and has contributed to the design of more flexible and powerful state of the art solution methodologies. Incorporating real life attributes into VRPs has created many variants of the problem, which fall in the following broader categories.

- Mixed VRPs: These problems are motivated by real operations and generally consist of a mix of existing VRP variants or extensions to the existing variants.
- Problem Specific: These are the VRPs which cannot be conformed to acronyms, as they contain specific constraints for a given industry or business. Once they are introduced to the literature, they are seldom revisited by other authors.
- Integrated problems: These problems also originate from industry operations where optimization is approached holistically and aspects such as routing, loading and production are solved simultaneously.
- VRPs with alternative objective: In some real applications the main objective of routing is not minimizing cost or time, but optimizing other aspects such as customer service, $CO_2$ emissions and sustainability factors, drivers working times, etc. These problems can be created with single or multi-objective framework, or even conflicting objective functions.

This chapter offers an evaluation of all categories of RVRPs, outlining some of the most recent applications and the methods used to solve the corresponding problems. We encourage the readers to investigate further the research we mention and gain more information on the specific mathematical formulation and methodology description. We also refer to useful review papers, which in some cases provide standard formulation and best performing methods for a given class of problems. Due to the problem-specific nature of RVRPs and the difficulty to summarize methodological aspects, we discuss some methodological and data considerations, which are not often highlighted in the literature.

The rest of the chapter is organized as follows. In Sect. 8.2 we present different definitions of RVRPs in the literature and offer our view on the important aspects of RVPRs. Section 8.3 contains a detailed literature review of the recent RVRP applications in the four main categories outlined above. Section 8.4 evaluates the main methodological issues that may arise when dealing with RVRPs, alongside directions for future research, where Sect. 8.5 concludes the chapter.

## 8.2   Background and Definition

The notion for real life problems is not new, because as early as 1993 there are papers in the literature, which were based on case studies or explicitly state that they were researching a real life problem. An example of that is the paper by Semet and Taillard [1], which tackles a real life VRP inspired by the grocery industry in Switzerland. Real life operations have been an inspiration for modelling in OR for the past 20 years. However, it was not until 2006 when real life problems were introduced as a class of the VRP. After the introduction of the Livestock Collection problem (LCP) [2], the authors formally categorize real life problems under the term 'rich' Vehicle Routing Problems (RVRP) and provide a loose definition of what a rich VRP is. Hasle et al. [3] state that a rich VRP includes aspects that are essential to the routing practice in real life and the richness of the problems can stem from many elements of the routing practice such as drivers, fleet, order types, depots, tours, etc. In addition, [4] states that rich VRPs should include different constraints or different objective functions. Vidal et al. [5] referred to the rich VRPs as Multi-Attribute VRPs (MAVRPs), which typically arise from real life situations. However, in their classification, any deviation from the Classical VRP is considered to be MAVRP. Goel and Gruhn [6] refer to the real life problem they propose as General VRP, whereas other authors do not adopt any classification terminology when dealing with these types of problems and simply refer to them as real life routing problems.

Caceres-Cruz et al. [7] state that RVRPs should incorporate relevant attributes of a real life vehicle routing distribution system, which may include dynamism, stochastic elements, heterogeneity, multi periodicity and many more. Lahyani et al. [8] also provide a comprehensive literature review of RVRPs as well as a theoretical framework. They investigate 41 publications since 2006, when the rich VRPs became a class of the VRP domain. The definition they provide states that a RVRP must have a sufficient number of real life routing elements in order to qualify as rich. It is mentioned that if the RVRP is defined in terms of physical characteristics, it should have at least nine additional elements to the classical VRP. If the problem is defined from a strategic or a tactical viewpoint, then it should have at least six additional elements. Rich VRP is a term that is becoming more popular in the literature, but there is no single definition of what exactly constitutes a rich VRP, which is accepted across the OR community. However, the argument on how much richness a problem should have in order to be categorised as rich can go beyond the number of features required in the nature of the problem. In our view, a RVRP should contain relevant aspects of a real life routing problem, which can aid the decision making process in practice and be significantly differentiable from other VRPs on the merits of their real life routing characteristics. In addition, when researching a real life VRP there should be some practical implications or recommendations for improving routing practice, which are relevant to a specific problem and ideally transferable to other applications. Working in collaboration with industry allows academics to measure the actual impact of their research, which can lead to interesting findings and contribution to our field.

## 8.3   Literature Review

Review papers on existing classes (variants) of VRP problems, such as the Fleet Size and Mix VRP (FSMVRP) or the VRP with Time Windows (VRPTW), typically summarize the most important aspects of the respective problems. Usually a standard formulation of the problem is provided alongside best known solutions on publicly available benchmark problem instances and outline of the best performing methods. However, RVRPs are very diverse and such a well-rounded summary could be very difficult to achieve. One of the main reasons for this is the loose definition provided for RVRPs. Another reason is that often a RVRP is introduced once and it is not revisited by another author under the same form with the same characteristics. Moreover, the nature of RVRPs changed over time and some of the problems referred to as rich in the past, may not necessarily qualify as ones today. RVRPs cannot be standardised, because their relevance and contribution stem from the diversity of real life routing practices.

We would like to note a few problem variants, which are not referred to as real life problems, but they are highly applicable in practice. The Two Echelon Vehicle Routing Problem (2E-VRP) considers a multilevel distribution system inspired by city logistics, where vehicles start at a depot, go to the nearest intermediate facility (satellite) and from there are routed to various customer locations. The purpose is to minimize pollution and congestion in big cities and avoid sending large trucks into the city. Instances with 21 customers are solved to optimality so far [9]. The Cumulative Vehicle Routing Problem (CumVRP) is motivated by customer satisfaction and relations. Its objective function is to minimize arrival times as opposed to cost. This is perhaps the most consumer centric variant in the VRP family, which incorporates issues like just in time service, equity and fairness [10] and [11]. The Multi-compartment Vehicle Routing Problem (MCVRP) arises when $m$ products must be delivered to customers by $k$ vehicles, which all have different compartments for each product. This variant considers the benefits of co-transportation as opposed to un-partitioned trucks and independent distribution. The Open Vehicle Routing Problem (OVRP) occurs when a given fleet of vehicles does not have to end the tour at the depot, but at the last customer. This variant is highly applicable for leased vehicles or any fleet that is not an asset of the company or in cases where drivers go home or another parking location after the tour is complete. Last, but not least is the Truck and Trailer Vehicle Routing Problem (TTRP) introduced by Chao [12]. It is inspired by the ability to access customer locations in difficult areas. TTRP consists of finding shortest routes to serve a set of customers either by the full vehicle (truck and trailer) or truck only.

### 8.3.1 Mixed Variants

It is rare in the industry that a fleet optimisation will be only constrained by time window alone, or heterogeneous fleet alone. Usually a combination of requirements and constraints is present, which need to be considered simultaneously. Therefore, academics become more creative and practical by introducing to the literature 'mixed' or 'extended' variants. An example of a mixed variant is a combination between VRPTW and VRP with Pickup and Delivery researched by Bent and van

Hentenryck [13]. Bektas and Laporte [14] combine VRP with Stochastic demand and time windows and [15] address a mixed variant between Multi Depot VRP and FSMVRP. Bortfeld [16] includes the presence of three-dimensional loading constraints, which makes the problem a mix of routing and loading decisions. Kang et al. [17] combine Multi Depot and Multi Period VRP, whereas more recently [18] combine Multiple Trip VRP with Backhauls. There are many other mixed variants, which exist in the literature and cannot be exhaustively listed, but are following the same principle. Extensions to the main variants are also common in the literature such as the VRPmiTW, which is VRPTW with multiple independent time windows [19] and the DVRP which is VRP with time dependent travel times [20].

## 8.3.2   Problem Specific

These category of problems are the RVRPs, which reflect the nature of real life operations. Not all papers we review in this chapter specifically state that the problem at hand is a 'rich' or a 'real life' problem, but they are inspired by industry operations and accommodate unique aspects and constraints. Martins et al. [21] introduce a Multi-compartment VRP, which has product-oriented time windows. This is relevant in the grocery industry, where the delivery time windows depend on the product type, whether it is fresh or frozen. Ren et al. [22] discuss the multishift VRP with overtime, which emphasizes on the drivers considerations and constraints, where [23] analyse the VRP with deadlines and travel/demand time, which is consumer centred. Archetti et al. [24] introduce a real life VRP with occasional drivers and solve it with a multi-start heuristic, while [25] introduce a RVRP with desynchronized arrivals to depot. Battarra et al. [26] solve the Multiple Minimum Trip VRP (MMTVRP), which is also industry inspired and assumes that one vehicle can be assigned to more than one route. Seixas and Mendes [27] incorporate drivers working hours into a Multiple Trip VRP with Heterogeneous Fleet. Franca et al. [28] consider a problem motivated by the selective waste collection of recyclables in Rio de Janeiro. The authors use the ORTEC routing software to test different scenarios for waste collection, where they explore the trade-offs between selective and non-selective waste collection. They found that even though some additional routes were generated and the total mileage increased, the cost of serving a client decreased by 28%, as well as the total overall cost. This is a very important finding, which can be used as a baseline for research in other countries.

Kramer et al. [29] consider a problem applied to the pharmaceutical distribution in Tuscany with auxiliary depots and anticipated deliveries. It is an interesting problem with various features such as incompatibilities between customers and routes and a maximum number of customers serviced. The authors solved this problem by a multi-start Iterated Local Search (ILS). Alcazar et al. [30] consider a rich VRP with last mile outsourcing decisions, driver hour regulations, incompatibility among goods, etc. The authors design a hybrid heuristic consisting of elements from Tabu Search (TS), Simulated Annealing (SA) and Variable Neighbourhood

Search (VNS), where they are specifically adapted to deal with the multiple real life constraints of the problem. Bianchessi et al. [31] consider a VRP with split delivery and time windows, where one of the aims is to limit customer inconvenience.

Tarantilis [32] uses a hybrid metaheuristic method based on TS, VNS and Guided Local Search (GLS) for the VRP with intermediate replenishment facilities. Liu and Jiang [33] use a memetic algorithm to solve the Closed-Open VRP, while [34] adopts a Column Generation based heuristic and a Greedy Randomized Adaptive Search Procedure (GRASP) for the min-max selective VRP. Benjamin and Beasley [35] propose a waste collection VRPTW with driver's rest period and multiple disposal facilities. They develop a hybrid metaheuristic method using TS and VNS, which performed well on selected benchmark problems. It is a common practice that authors often create or adapt methodologies to fit the richness of the problem they are investigating. For instance, [22] solved the VRP with multi shift and overtime and introduced a shift-dependant heuristic to tackle the problem. Some problems are motivated by urban logistics and are greatly applicable in a city setting. For instance, recently [36] propose a complex 2E-VRP with multi-objective function, vehicle synchronization and the notion of grey zone customers. It examines the assignment of those customers which are at the boundary of the two echelons and explores different city layouts. Bevilaqua et al. [37] also discuss a 2E-VRP, with heterogeneous fleet and real life constraints, motivated by Brasilian wholesale companies.

Simeonova et al. [38] introduce a RVRP with heterogeneous fleet, light load customers and overtime, motivated by the gas delivery industry. The authors solve the problem using a Population VNS with Adaptive Memory. Mancini [39] introduces and formalizes a problem with multi depot, multi period and heterogeneous fleet and solved it using an Adaptive Large Neighbourhood Search (ALNS) heuristic. Though it is not specifically applied to a given company, it is a problem that can arise in industry. Both papers provide a MIP formulation and a heuristic method to solve the problem, as well as a comparison of the exact method to the performance of the heuristic algorithm. This is a very good practice when it comes to RVRPs, because there are no available literature benchmark instances for RVRPs.

### 8.3.3 Problems with Alternative Objective Function

Some problems have objective function alternative to minimize cost, such as minimize carbon emissions or maximize value. Other problems can be multi-objective, where there is multiple criteria for decision making. The objectives can also be conflicting and there is no single solution which simultaneously optimizes all criteria. Therefore, we look for Pareto optimal solutions (also known as non-dominated solutions) and analyse the trade-offs between the different criteria. Spliet and Gabor [40] address a problem where the objective is to have consistent time windows which are assigned to customers before their demand is known. Kovac et al. [41] solve the Consistent VRP (ConVRP) where customers should be visited by the same driver and deviations from a given arrival time are penalized. Papers with similar objective show the clear trade-off between routing cost and service consistency, and propose ways on how to improve customer service at a minimum increase in cost. The authors have estimated that the service consistency can be improved by 70% with only 4% increase in cost.

Rodriguez-Martin et al. [42] also consider driver consistency in a context of Periodic VRP, which is aimed at improving customer service. Similarly to [41], the authors report that the cost of consistency is 4% on average. There are examples in industry where this is required, because it improves the driver's learning. In some cases drivers also do other activities such as special loading requirements, shelf replenishment, etc. However, having the same driver visit the same customers also contributes to other aspects which are not greatly discussed in the literature. Drivers prefer to have the same delivery areas every day, because they feel more comfortable with getting to the customer's locations and are familiar with the nature of the delivery. This helps them to be more efficient and could contribute to reducing the service time and the driving time to the customer. Hoogeboom and Dullaert [43] solve the VRP with arrival time diversification. This is a very interesting problem, which is relevant for the Cash-in-Transit (CIT) companies, which deliver valuable goods. The objective here is not only to minimize cost but also to create an unpredictable route, with alternating arrival times at the customers, which is important for security reasons.

We mentioned in the introduction that the development of the VRP field is greatly inspired by changes in industry and policy. Companies and governments are investing in more environmentally friendly vehicles because of the increased global awareness and necessity to lower our carbon footprint. Therefore, there is emerging research around Green VRPs and Electric VRPs, where the aim is to minimize carbon emissions and environmental footprint. The literature on the green road transportation has mainly concentrated on the issue of emissions and [44] were one of the first to highlight it. Bektas and Laporte [14] presented a study based on the classical Vehicle Routing Problem (VRP), called Pollution-Routing Problem (PRP) optimizing travel distance and greenhouse emissions. Wassan et al. [45] provide an extensive summary on Green Reverse logistics, which shows the benefit of incorporating green aspects in delivery and collection problems. Poonthalir and Nadarajan [46] conducted research on fuel consumption and concluded that greater efficiency can be achieved under varying speed limit. Fathollahi-Fard et al. [47] address a Green Home Health Care (GHHC) Supply chain problem, as a bi-objective location-allocation-routing problem with the aim of achieving sustainability for HHC.

Electric vehicles (EVs) are by definition more energy efficient, but they come with different constraints such as their range is limited by the battery life. The objective function is typically to minimize distance or cost, but also to optimize the battery life, as well as the optimal location for the recharging facilities. Some studies consider hybrid vehicles, where the battery life is firstly used up, and then the vehicle switches to regular fuel. Badin et al. [48] presented a study that quantifies the influence on energy consumption of different factors such as driving conditions, auxiliaries' impact, driver's aggressiveness and braking energy recovery strategy on an electric vehicle. Wua et al. [49] proposed an analytical electric vehicle power estimation, which concludes that EVs are more efficient when driving on in-city routes rather than motorway routes. Lin et al. [50] studied a general Electric Vehicle Routing Problem (EVRP) which considers the load effect on battery consumption. Hiermann et al. [51] introduce an Electric Fleet Size and Mix Vehicle Routing Problem with Time Windows and Recharging Stations (E-FSMFTW), where they optimize the minimum cost routing and fleet composition (a mix of electric and conventional vehicles), but also the recharging times and locations. Macrina et al. [52] introduce a complex energy-efficient Green VRP, with a mix of conventional and electric vehicles, partial battery recharging and incorporating the effect of the braking regenerating system. All these studies can help society and businesses with the global transition to more energy-efficient travel and highlight any possible issues and ways to overcome them.

### 8.3.4  Integrated Problems

The integrated problems are another interesting area with real life application. Recently there is an increasing number of publications which tackle VRPs in combination with other aspects such as loading, inventory, machine scheduling, etc. These problems are quite important as they show the potential savings and improvement in operations and supply chain, when we approach problems in an integrated manner rather than sequentially. Van Gils et al. [53] formulate and solve a complex integrated problem, which includes order batching and picker routing and scheduling. The problem also has real life aspects such as high-level locations, order due times and limited availability of pickers, where the objective is to increase order picking efficiency. The authors used ILS to solve the problem and found substantial savings of 16.9% from using this integrated approach, applied to a real case study. Hoogeboom and Dullaert [43] research a deteriorating inventory routing problem applied in the liquefied natural gas distribution network. The authors use ALSN and provide managerial insights regarding alternative replenishment strategies, some interesting trade-offs and observations, and different effects on the problem such as gas evaporation rates. Bertazzi et al. [54] also consider inventory routing, but with multiple depots and solves the problem using a three-phase matheuristic method, consisting of clustering, routing and optimization phase. Liu et al. [55] also examine an interesting real life inventory routing problem, which also has an alternative objective function which relates to the maximization of the service levels under budget limitations.

Integrating routing with location is also an important problem with application in real life. Darvish et al. [56] solve the flexible two-echelon location-routing problem and the authors find average savings from the integration of up to 30% for businesses. Ghaderi and Burdett [57] consider location routing for hazardous materials in a bi-modal transportation network, which aims to minimize cost and risk. Zheng et al. [58] integrate routing, location and inventory aspects in a supply chain network design with real life constraints, which they solve with an exact method. It is worth noting that there is practical and controversial issue in combining these two problems as these have different time frame, where location is usually a strategic problem, whereas routing is operational. This issue was firstly investigated by Salhi and Nagy [59]. The reader is referred to [60] and [61] for surveys on location-routing problems. Routing can also be integrated with machine scheduling, which can assist companies with supply chain flow optimization and management. For instance, [62] aim to minimize carbon emissions for an integrated routing and machine scheduling problem, which can provide a greener approach to manufacturing practices.

## 8.4 Methodological Considerations

The papers we evaluated in this chapter are examples on the need to adjust, extend and be creative with existing methodologies in order to make them applicable to real life problems and address their special features. All those real life aspects that can be added to a VRP problem and provide that extra richness contribute to making RVRP research very flexible and a fertile area for ideas and novel developments. However, there is an issue which has not been specifically addressed in the literature, which relates to the solution methodologies. In the VRP family, a contribution to the literature is mainly considered if a new interesting problem is introduced, which is different from previous research or a new methodology, which is powerful and relevant in terms of performance and novelty. However, given the fact that RVRPs are so different from other variants and from each other, it poses a challenge for proving algorithmic efficiency compared to other methods in the literature. Table 8.1 shows a summary of the most recent applications reviewed in the chapter so far, outlining their methodological choices. We present papers from each of the problem categories discussed earlier, the nature of the data used in the paper, whether there is comparability of the results and whether the method has been implemented in a real life setting with real savings estimation. Column 3 in the table refers to datasets which are either randomly generated or hypothetical, which simulate real life operations. Column 4 refers to adaptations from literature benchmark instances, where column 5 shows if the research is tested on data provided by industry. Comparability includes any approach to results validation, such as comparing against Best Known Solutions (BKS), against results from an exact method (or the main methodology of the paper is exact), or against baseline results generated by a company.

Looking at Table 8.1, we can make a few observations. There are very few applications, which are compared against BKS or extend their methodologies to other problems. Forty-five percent of the cited papers test their real life problems on actual data provided by businesses, but none is implemented and used by those businesses. Even though the authors make highly relevant and useful recommendations and provide managerial insight, actual implementation can help understand the true value of optimization and how much of the projected benefits can be materialized.

**Table 8.1** Methodological review of recent real life applications

| Paper | Category | Generated dataset | Adapted dataset | Real dataset | Comparability | Implementation |
|---|---|---|---|---|---|---|
| Alcazar et al. [30] | Problem specific | ✗ | ✓ | ✓ | ✗ | ✗ |
| Anderluh et al. [36] | Alternative objective | ✓ | ✓ | ✓ | ✗ | ✗ |
| Archetti et al. [24] | Problem specific | ✓ | ✗ | ✗ | ✗ | ✗ |
| Badin et al. [48][a] | Integrated problem | ✗ | ✓ | ✓ | ✓ | ✗ |
| Bertazzi et al. [54][b] | Problem specific | ✗ | ✓ | ✓ | ✗ | ✗ |
| Bevilaqua et al. [37] | Integrated problem | ✓ | ✗ | ✗ | ✗ | ✗ |
| Darvish et al. [56][c] | Problem specific | ✗ | ✗ | ✓ | ✗ | ✗ |
| Franca et al. [28] | Integrated problem | ✓ | ✗ | ✗ | ✗ | ✗ |
| Ghaderi and Burdett [57][a] | Integrated problem | ✗ | ✓ | ✗ | ✓ | ✗ |
| Ghiami et al. [63][b] | Problem specific | ✗ | ✓ | ✓ | ✗ | ✗ |
| Hoogeboom and Dullaert [43] | Problem specific | ✗ | ✓ | ✗ | ✗ | ✗ |
| Kovac et al. [41] | Problem specific | ✗ | ✓ | ✓ | ✗ | ✗ |
| Kramer et al. [29] | Problem specific | ✗ | ✓ | ✓ | ✗ | ✗ |
| Lee and Lee [23][a] | Mixed problem | ✗ | ✓ | ✗ | ✓ | ✗ |
| Mancini [39] | Mixed problem | ✗ | ✓ | ✗ | ✓ | ✗ |
| Martins et al. [21] | Problem specific | ✗ | ✗ | ✓ | ✓ | ✗ |
| Naji-Azimi et al. [25] | Problem specific | ✗ | ✓ | ✗ | ✓ | ✗ |
| Rodriguez-Martin et al. [42] | Problem specific | ✗ | ✓ | ✗ | ✓ | ✗ |
| Simeonova et al. [38][d] | Integrated problem | ✗ | ✗ | ✓ | ✓ | ✗ |
| Van Gils et al. [53] | Mixed problem | ✗ | ✓ | ✗ | ✓ | ✗ |
| Wassan et al. [18] | Integrated problem | ✗ | ✗ | ✓ | ✓ | ✗ |
| Zheng et al. [58] | Alternative objective | ✗ | ✓ | ✗ | ✓ | ✗ |
| Fathollahi-Fard et al. [47] | Alternative objective | ✓ | ✗ | ✗ | ✗ | ✗ |
| Macrina et al. [52] | Alternative objective | ✗ | ✓ | ✗ | ✗ | ✗ |
| Bianchessi et al. [31] | Problem specific | ✓ | ✗ | ✗ | ✗ | ✗ |

[a] Compared against exact method
[b] Compared against BKS
[c] The authors used routing software to solve the problem
[d] Compared against real operations

## 8.4.1  Methodological Comparability

Real life problems have unique characteristics and additional requirements, which lead to extra restrictions. If we consider a classical mathematical formulation (ILP), exact methods generally find it hard to solve RVRPs, as the problem becomes too large and very small instances are solved to optimality. Column generation, however, seems to perform better on RVRPs, because the extra constraints allow for the elimination of columns, which can decrease the complexity. This could be one of the reasons why most exact methods for RVRPs focus on CG and Branch-Cut-Price. For instance, [4] used Branch-and-Cut for Rich VRP with docking constraints. Dayarian et al. [64] developed a Column Generation (CG) method for a real life case inspired by milk collection, where [65] applied CG to the livestock collection problem.

Heuristic methods are more flexible and adaptive in their design, therefore adding special features can actually speed up their performance. This makes them a preferred option when solving RVRPs, but they also have some limitations. They are generally problem specific and cannot guarantee optimality. Coupled up with the problem-specific nature of the RVRPs it may raise a question for methodological justification. This is the main reason why we need consistent and reliable approach to show the algorithmic strength and adaptability of the methods we create in the RVRP domain, which would strengthen their contribution. In fact, researching a real life variant can act as an inspiration to adapt and adjust well known methods and provide an opportunity to extend those methods to other problems and ideally make them more generalizable across VRP problems.

We already mentioned an example of good practice when it comes to demonstrating algorithmic efficiency, namely generating optimal solutions and LB/UB where possible and compare against the heuristic solution using commercial solvers such as CPLEX and Gurobi. The larger the instances are, the gap between the bounds is more likely to increase, though not always the case. However, having the heuristic solution within those bounds could be a good indication for the quality and appropriateness of the heuristic method.

Another suggestion is to reduce the RVRP to a well known and researched VRP variant, which has publicly available benchmark instance sets. If the RVRP is quite unique, most likely it could be reduced to the capacitated VRP (CVRP) or some of the most common problems such as VRPTW, VRPPD and FSMVRP. In this scenario the method is not expected to outperform existing BKS, because it is created for a real life problem with unique constrains. However, if the method performs reasonably well and fast on the standard benchmark instances, with relatively small gaps from BKS, it is an indication that the methods are powerful. One of the trends in the literature is to design more generalizable algorithms, which can be applied to a range of VRPs. Testing an algorithm which is specifically designed for a given RVRP on different instance sets shows not only algorithmic efficiency but also generalizability and adaptability.

Regarding preferred methods in the literature for RVRPs, there is no one best method, because different aspects of the heuristic methods fit different RVRPs.

However, it seems that the ILS and ALNS are getting more attention, especially when it comes to problems with real life constraints. One of the main reasons could be that ALNS is equipped with destroy and repair mechanisms which can be adapted to fit special features. For instance, [21] recently use daily and weekly operators within their method to reflect the nature of their RVRP.

## 8.4.2   Data Considerations

Following the methodological discussion, there is also the question on what data we should use when we address RVRPs, as algorithm testing and data are closely related. There are a few good practices regarding data which we would like to outline. Some papers on RVRPs are case studies, based on real company data. Using a real dataset, which is also used by the company in question, is a good way of showing algorithmic efficiency by directly comparing the results from the study to the actual practices of the company. By doing this, the impact of the study can be measured and recommendations can be made on how to improve the routing practice. Moreover, using real data can impact the behaviour of the algorithm and lead to very interesting practical insights, which is very important when addressing RVRPs.

In other cases, the datasets used for the RVRPs can be either random or generated in a way to simulate real operations. In these cases the issue of comparability is more critical. Therefore, it is important that some form of comparison or test of algorithmic merits is adopted, or some practical applicability is demonstrated when the algorithm is used by the business which motivated the research. Some authors test their methodology on adapted benchmark instances from the literature, which are well known and available. In this case we suggest that the authors run their algorithms on the original instances they were adapted from, to demonstrate their flexibility and capability to address similar problems.

It has to be noted here that some papers propose algorithms for RVRPs which aim to be all-encompassing rich solvers, rather than single methodologies for one particular problem. An example of this is the Genetic Algorithm based rich solver proposed by Vidal et al. [5]. In their paper many aspects of the VRP, including various rich elements can be addressed by the proposed methodology, hence the results can be tested on many literature benchmarks. However, the purpose of the method is to be all-encompassing and able to accommodate VRPs across the different variant classes.

## 8.4.3   Implementation considerations

As demonstrated in Table 8.1, not all research in the RVRP domain is tested on datasets provided by industry. This leads to a very interesting question of practical applicability, implementation and the real use of real life routing methodologies in practice. Recently there have been some discussions on international forums and conferences about the implementation issues of OR models and methods in business operations [66], and a realisation that very often the methodologies developed to improve company operations do not result in the desired and estimated benefits. There are many barriers to implementation and adoption of new methodologies, such as resistance to change, long standing perceptions, even personal preferences of drivers and other personnel. We suggest that authors aim to incorporate their findings and recommendations into the daily decision making of the companies the research is motivated by. By doing so, we can observe the true impact of optimization and how much of the savings we compute are actually realized. Moreover, we can discover how well our methodologies fit into the daily decision making routines and if there are any aspects of the organisational structure which can strengthen or compromise the desired results. For instance, [67] state that visual attractiveness of the vehicle routes can assist implementation, because they appear more user-friendly to the scheduling staff. Therefore, generating routing schedules which are compact, not overlapping and not complex can increase the adoption of routing algorithms in practice. Academics should recognise the issue of implementation, especially in the face of current literature trends to bring academia closer to industry. Moreover, there are numerous relevant and important real life applications which we discussed throughout the chapter and if they are properly implemented by businesses and governments can result in significant financial, social and environmental gains.

## 8.5   Conclusion

In this chapter, we presented the recent developments and applications of real life vehicle routing problems. We discussed the origins and definition of RVRPs and how they differ from the other problems in the VRP domain. In particular, we reviewed Mixed VRPs, Problem-Specific VRPs with unique real life features, Integrated problems and those with Alternative objectives, including green and sustainable practices. We also discussed some important methodological considerations regarding algorithm comparability and generalizability. Some important data-related challenges are outlined when dealing with RVRPs, alongside suggestions on how to overcome them and make a stronger contribution to the literature and routing practice. Lastly, we touched on the importance of implementation when it comes to RVRPs. By definition they are motivated by real operations and should result in practical implications and managerial insight. If we ensure the successful implementation of VRPs in general, we can see the full potential of optimization and the actual benefit realisation for businesses. More importantly, a lot of the RVRP research is aimed not only to achieve economic gains but also to reduce the carbon footprint of routing operations, to maximize the added value for customers and improved experience for the society. Benefit actualisation through implementation of real life problems can achieve our quest for minimizing the gap between academia and industry, which will make our findings not only academically viable but also an important tool for business and governments to build a more efficient and greener economy.

**Acknowledgments** The authors would like to thank the referees for their valuable suggestions, which improved the content and presentation of the chapter.

# References

1. Semet, F., Taillard, E.: Solving real-life vehicle routing problems efficiently using Tabu Search. Ann. Oper. Res. **41**, 469–488 (1993)
2. Gribkovskaia, I., Gullberg, B.O., Hovden, K.J., Wallace, S.W.: Optimization model for a livestock collection problem. Int. J. Phys. Distrib. Logist. Manag. **36**, 136–152 (2006)
3. Hasle, G., Løkketangen, A., Martello, S.: Rich models in discrete optimization: Formulation and resolution. Eur. J. Oper. Res. **175**, 1752–1753 (2006)
4. Rieck, J., Zimmermann, J.: A new mixed integer linear model for a rich vehicle routing problem with docking constraints. Ann. Oper. Res. **181**, 337–358 (2010)
5. Vidal, T., Crainic, T., Gendreau, M., Prins, C.: A unified solution framework for multi-attribute vehicle routing problems. Eur. J. Oper. Res. **234**, 658–673 (2014)
6. Goel, A., Gruhn, V.: A general vehicle routing problem. Eur. J. Oper. Res. **191**, 650–660 (2008)
7. Caceres-Cruz, J., Arias, P., Guimarans, D., Riera, D., Juan, A.: Rich vehicle routing problem: survey. ACM Comput. Surv. **47**, 1–28 (2014)
8. Lahyani, R., Khemakhem, M., Semet, F.: Rich vehicle routing problems: From a taxonomy to a definition. Eur. J. Oper. Res. **241**, 1–14 (2015)
9. Hemmelmayr, V.: An adaptive large neighbourhood search heuristic for two-echelon vehicle routing problems arising in city logistics. Comput. Oper. Res. **39**, 3215–3228 (2012)
10. Ngueven, S., Prins, C., Calvo, R.: An effective memetic algorithm for the cumulative capacitated vehicle routing problem. Comput. Oper. Res. **37**, 1877–1885 (2010)
11. Sze, J.E., Salhi, S., Wassan, N.: The cumulative capacitated vehicle routing problem with min-sum and min-max objectives: An effective hybridisation of adaptive variable neighbourhood search and large neighbourhood search. Transp. Res. B **101**, 162–184 (2017)
12. Chao, I.M.: A tabu search method for the truck and trailer routing problem. Comput. Oper. Res. **29**, 33–51 (2002)
13. Bent, R., van Hentenryck, P.: A two-stage hybrid algorithm for pickup and delivery vehicle routing problems with time windows. Comput. Oper. Res. **33**, 875–893 (2006)
14. Bektas, T., Laporte, G.: The pollution-routing problem. Transp. Res. B **45**, 1232–1250 (2011)
15. Salhi, S., Imran, A., Wassan, N.: The multi-depot vehicle routing problem with heterogeneous vehicle fleet: formulation and a variable neighbourhood search implementation. Comput. Oper. Res. **52**, 315–325 (2014)
16. Bortfeld, A.: A hybrid algorithm for the capacitated vehicle routing problem with three-dimensional loading constraints. Comput. Oper. Res. **39**, 2248–2257 (2010)
17. Kang, K., Lee, Y., Lee, B.: An exact algorithm for multi depot and multi period vehicle scheduling problem. In: Computer Science and Applications, pp. 350–359 (2005)
18. Wassan, N., Wassan, N.A., Nagy, G., Salhi, S.: The multiple trip vehicle routing problem with Backhauls: formulation and a two-level variable neighbourhood search. Comput. Oper. Res. **78**, 454–467 (2017)
19. Doerner, K., Gronald, M., Hartl, R., Kiechle, G., Reimann, M.: Exact and heuristic algorithms for the vehicle routing problem with multiple interdependent time windows. Comput. Oper. Res. **35**, 3034–3048 (2008)

20. Haghani, A., Jung, S.: A dynamic vehicle routing problem with time-dependent travel times. Comput. Oper. Res. **32**, 2959–2986 (2005)
21. Martins, S., Ostermeier, M., Amorim, P., Huebner, A., Almada-Lobo, B.: Product-oriented time window assignment for a multi-compartment vehicle routing problem. Eur. J. Oper. Res. **276**, 893–909 (2019)
22. Ren, Y., Dessouky, M., Ordonez, F.: The multi-shift vehicle routing problem with overtime. Comput. Oper. Res. **37**, 1987–1998 (2010)
23. Lee, C., Lee, K.: Robust vehicle routing problem with deadlines and travel time/demand uncertainty. J. Oper. Res. Soc. **63**, 1294–1306 (2012)
24. Archetti, C., Savelsbergh, M., Speranza, M.: The vehicle routing problem with occasional drivers. Eur. J. Oper. Res. **254**, 472–480 (2016)
25. Naji-Azimi, Z., Salari, M., Renaud, J., Ruiz, A.: A practical vehicle routing problem with desynchronized arrivals to depot. Eur. J. Oper. Res. **255**, 58–67 (2016)
26. Battarra, M., Monaci, M., Vigo, D.: An adaptive guidance approach for the heuristic solution of a minimum multiple trip vehicle routing problem. Comput. Oper. Res. **36**, 3041–3050 (2009)
27. Seixas, M.P., Mendes, A.B.: Column generation for a multitrip vehicle routing problem with time-windows, driver work hours, and heterogeneous fleet. Math. Probl. Eng. **8**, 1–13 (2013)
28. Franca, L.S., Ribeiro, G.M., Chaves, G.L.D.: The planning of selective collection in a real-life vehicle routing problem: A case in Rio de Janeiro. Sustain. Cities Soc. **47**, 101–488 (2019)
29. Kramer, R., Cordeau, J.F., Iori, M.: Rich vehicle routing with auxiliary depots and anticipated deliveries: An application to pharmaceutical distribution. Transp. Res. E **129**, 162–174 (2019)
30. Alcazar, J., Caballero-Arnaldos, L., Vales-Alonso, J.: Rich vehicle routing problem with last-mile outsourcing decisions. Transp. Res. E **129**, 263–286 (2019)
31. Bianchessi, N., Drexl, M., Irnich, S.: The split delivery vehicle routing problem with time windows and customer inconvenience constraints.Transp. Sci. **53**, 1067–1084 (2019)
32. Tarantilis, C.: A hybrid guided local search for the vehicle-routing problem with intermediate replenishment facilities. INFORMS J. Comput. **20**, 154–168 (2008)
33. Liu, R., Jiang, Z.: The close-open mixed vehicle routing problem. Eur. J. Oper. Res. **220**, 349–360 (2012)
34. Valle, C.: Heuristic and exact algorithms for a min–max selective vehicle routing problem. Comput. Oper. Res. **38**, 1054–1065 (2011)
35. Benjamin, A.M., Beasley, J.E.: Metaheuristics for the waste collection vehicle routing problem with time windows, driver rest period and multiple disposal facilities. Comput. Oper. Res. **37**, 2270–2280 (2010)
36. Anderluh, A., Nolz, P., Hemmelmayr, V., Crainic, T.: Multi-objective optimization of a two-echelon vehicle routing problem with vehicle synchronization and 'grey one' customers arising in urban logistics. Eur. J. Oper. Res. (2019). https://doi.org/10.1016/j.ejor.2019.07.049
37. Bevilaqua, A., Bevilaqua, D., Yamanaka, K.: Parallel island based memetic algorithm with Lin–Kernighan local search for a real-life two-echelon heterogeneous vehicle routing problem based on Brazilian wholesale companies. Appl. Soft Comput. J. **76**, 697–711 (2019)
38. Simeonova, L., Wassan, N., Salhi, S., Nagy, G.: The heterogeneous fleet vehicle routing problem with light loads and overtime: Formulation and population variable neighbourhood search with adaptive memory. Expert Syst. Appl. **114**, 183–195 (2018)
39. Mancini, S.: A real-life multi depot multi period vehicle routing problem with heterogeneous fleet: Formulation and adaptive large neighbourhood search based metaheuristic. Transp. Res. C **70**, 100–112 (2016)
40. Spliet, R., Gabor, A.: The time window assignment vehicle routing problem. Transp. Sci. **49**, 721–731 (2015)
41. Kovac, A., Parragh, S., Hartl, R.: The multi-objective generalized consistent vehicle touring problem. Eur. J. Oper. Res. **247**, 441–458 (2015)
42. Rodriguez-Martin, I., Salazar-Gonzalez, J.J., Yaman, H.: The periodic vehicle routing problem with driver consistency. Eur. J. Oper. Res. **273**, 575–584 (2019)
43. Hoogeboom, M., Dullaert, W.: Vehicle routing with arrival time diversification. Eur. J. Oper. Res. **275**, 93–107 (2019)

44. McKinnon, A.C., Piecyk, M.I.: Measurement of CO2 emissions from road freight transport: A review of UK experience. Energy Policy **37**, 3733–3742 (2009)
45. Wassan, N., Wassan, N., Simeonova, L ., Besbes, W.: Green reverse logistics: case of the vehicle routing problem with delivery and collection demands. Solving Transport Problems: Towards Green Logistics. In this pp. 161–183. Wiley (2019)
46. Poonthalir, G., Nadarajan, R.: A fuel efficient green vehicle routing problem with varying speed constraint (F-GVRP). Expert Syst. Appl. **100**, 131–144 (2018)
47. Fathollahi-Fard, A.M., Govindan, K., Hajiaghaei-Keshteli, M., Ahmadi, A.: A green home health care supply chain: New modified simulated annealing algorithms. J. Cleaner Prod. **10**, 118200 (2019)
48. Badin, F., Le Berr, F., Briki, H., Dabadie, J.-C., Petit, M., Magand, S., Condemine, E.: Evaluation of EVs energy consumption influencing factors, driving conditions, auxiliaries use, driver's aggressiveness. World Electr. Veh. J. **6**, 2032–6653 (2013)
49. Wua, X., Freese, D., Cabrera, A., Kitch, W.A.: Electric vehicles energy consumption measurement and estimation. Transp. Res. D **34**, 52–67 (2015)
50. Lin, J., Zhou, W., Wolfson, O.: Electric vehicle routing problem. Transp. Res. Procedia **12**, 508–521 (2016)
51. Hiermann, G., Puchinger, J., Ropke, S., Hartl, R.: The electric fleet size and mix vehicle routing problem with time windows and recharging stations. Eur. J. Oper. Res. **252**, 995–1018 (2016)
52. Macrina, G., Laporte, G., Guerriero, F., Di Puglia Pugliese, L.: An energy-efficient green-vehicle routing problem with mixed vehicle fleet, partial battery recharging and time windows. Eur. J. Oper. Res. **276**, 971–982 (2019)
53. Van Gils, T., Ramaekers, K., Braekers, K., Depaire, B., Caris, A.: Increasing order picking efficiency by integrating storage, batching, zone picking and routing policy decision. Int. J. Prod. Econ. **197**, 243–261 (2019)
54. Bertazzi, L., Coelho, L.C., De Maio, A., Lagana, D.: A metaheuristic algorithm for the multi-depot inventory routing problem. Transp. Res. E **122**, 524–544 (2019)
55. Liu, M., Liu, X., Chu, F., Zheng, F., Chu, C.: Distributionally robust inventory routing problem to maximize the service level under limited budget. Transp. Res. E **126**, 190–211 (2019)
56. Darvish, M., Archetti, C., Coelho, L., Speranza, M.: Flexible two-echelon location routing problem. Eur. J. Oper. Res. **277**, 1124–1136 (2019)
57. Ghaderi, A., Burdett, R.: An integrated location and routing approach to transporting hazardous materials in a bi-modal transportation network. Transp. Res. E **127**, 49–65 (2019)
58. Zheng, X., Yin, M., Zhang, Y.: Integrated optimization of location, inventory and routing in supply chain network design. Transp. Res. B **121**, 1–20 (2019)
59. Salhi, S., Nagy, G.: Consistency and robustness in location-routing. Stud. Locat. Anal. **13**, 3–19 (1999)
60. Salhi, S., Nagy, G.: Location-routing: Issues, models and methods. Eur. J. Oper. Res. **177**, 649–672 (2007)
61. Schneider, M., Drexl, M.: A survey of variants and extension of the location-routing problem. Eur. J. Oper. Res. **241**, 283–308 (2015)
62. Wang, J., Yao, S., Sheng, J., Yang, H.: Minimizing total carbon emissions in an integrated machine scheduling and vehicle routing problem. J. Cleaner Prod. **229**, 1004–1017 (2019)
63. Ghiami, Y., Demir, E., Van Woensel, T., Christiansen, M., Laporte, G.: A deteriorating inventory routing problem for an inland liquefied natural gas distribution network. Transp. Res. B **126**, 45–67 (2019)
64. Dayarian, I., Crainic, T., Gendreau, M., Rej, W.: A column generation approach for a milti-attribute vehicle routing problem. Eur. J. Oper. Res. **241**, 888–906 (2015)
65. Oppen, J.: Algorithmic and computational methods in retrial queues solving a rich vehicle routing and inventory problem using column generation. Comput. Oper. Res. **37**, 1308–1317 (2010)
66. Gromichio, J.: How appealing is an optimum? In: Verolog Conference, Seville 02.06.2019–05.06.2019 (2019)
67. Rossit, D., Vigo, D., Tohme, F., Frutos, M.: Visual attractiveness in routing problems: A review. Comput. Oper. Res. **103**, 13–34 (2019)

Printed in the United States
by Baker & Taylor Publisher Services